The Electric Car Guide

2016 Edition

Your Guide to Buying and Owning an Electric Car

Michael Boxwell

Foreword by Robert Llewellyn

Greenstream Publishing Limited
5 Palmer House
Palmer Lane
Coventry
CV1 1FN
United Kingdom

www.greenstreampublishing.com

Published by Greenstream Publishing 2016
Copyright © Michael Boxwell 2010–2016
First Edition – published February 2010
Second Edition *(Revised)* – published April 2010
Third Edition *(Revised)* – published February 2011
Fourth Edition *(Revised)* – published April 2012
Fifth Edition *(Revised)* – published November 2014
Sixth Edition *(Revised)* – published April 2016

ISBN 978-1-907670-60-2

Michael Boxwell asserts the moral right to be identified as the author of this work.

A catalogue record for this book is available from the British Library.

While we have tried to ensure the accuracy of the contents of this book, the author or publishers cannot be held responsible for any errors or omissions found therein.

All rights reserved. No part of this publication may be reproduced, stored in a retrieval system, or transmitted, in any form or by any means, electronic, mechanical, photocopying, recording or otherwise, without the prior permission of the publishers.

Table of Contents

About This Book 1

Foreword by Robert Llewellyn 3

Introducing the Electric Car 5
    What is an electric car? 7
    How an electric car compares to a conventional car 8
    Range 9
    Charging it up 11
    Travelling longer distances by electric car 12
    Charging at work 12
    Are electric cars the silent killer? 13
    Other forms of electric cars 15
    Chapter summary 22

Why are Electric Cars Important Today? 23
    The changing face of transport 23
    Energy security and price volatility 24
    Peak oil 25
    The multi-car family 27
    The generation gap 27
    Pollution 29
    The rise of alternative electric vehicles 29
    Chapter summary 32

Living with an Electric Car 33
    First impressions 33
    Plugging it in 36
    The first few weeks 37
    Borrowing a 'plug full' of electricity 40
    Electric car charging-points 42
    Long-distance driving 42
    Long term ownership 44
    Chapter summary 45

Will an Electric Car Work for Me? 47
    Range – the true story 50
    Where can I plug in? 52
    Can I charge my car elsewhere? 57
    Can I remove the batteries to recharge them? 57
    How often do I need to be able to drive further than an electric car will allow me to go? 58
    What if an electric car is not suitable for me now? 59
    Chapter summary 60

Purchasing and Running Costs 61

Leasing schemes and purchasing plans 61
Reaching the Tipping Point 62
Government subsidies and discounts 62
Buying second hand 63
Fuel costs 64
Servicing costs 65
Chapter summary 66

Electric Car Charging Networks 67
Charging-point standards 69
What cars work with what charging-points? 74
How to use a public charging-point 75
Charging-point networks in the United States 76
Charging-point networks in the United Kingdom 76
Chapter summary 77

Electric Car Categories 78
Categories of electric vehicle 78
Why are these classifications important? 83
Chapter summary 84

Electric Cars You Can Buy Today 85
Understanding the specifications 85
BMW i3 87
Chevrolet Spark EV 89
Citroen C-Zero / Mitsubishi i-MiEV / Peugeot iOn 91
FIAT 500e 92
Ford Focus Electric 93
KIA Soul EV 94
Mahindra e2o 95
Mercedes B-Class Electric Drive 96
Nissan e-NV200 97
Nissan LEAF 98
Renault Kangoo ZE 100
Renault Twizy 101
Renault Zoe 102
Tesla Model S 103
Tesla Model X 105
Tesla Model III 106
Tesla Roadster 108
VW e-Golf 109
VW e-Up! 109

Me and My Electric Car 110
Neil Butcher – Mitsubishi i-MiEV   United Kingdom 110
Peter Berg – Nissan LEAF   California, USA 111
Paul Edgington – Renault Fluence ZE   United Kingdom 114
Zarla Harriman – G-Wiz *i*   United Kingdom 115
'ECarFan' – Tesla Model S   California 116

    Martin Ingles – Mitsubishi i-MiEV   United Kingdom 118
    Michael Walsh – Nissan LEAF   California, USA 120
    Angela Boxwell – Mitsubishi i-MiEV   United Kingdom 121
    Damian Powell – Nissan LEAF   United Kingdom 122

Electric Cars and the Environment 124
    Before we begin 125
    Comparing electric cars with conventional cars 125
    An acknowledgement to motor manufacturers 127
    Air pollution 128
    Air pollution from electricity generation 137
    How our energy is produced 142
    Electric cars and electricity supply 168
    Fuel economy... 170
    The environmental impact of batteries 176
    The environmental impact of vehicle manufacturing and distribution......................... 179
    Vehicle recycling 181
    Chapter summary 182

Real World Economy Figures for Electric Cars... 184
    The test... 184
    The carbon calculations 185
    Test validity 185
    The electric cars.. 186
    The combustion engine cars 187
    Test results from the electric cars: 188
    What if they were powered by coal? 189
    How little electricity an electric car uses 189
    Test results from the combustion engine cars 190
    Side by side analysis: well-to-wheel measurements... 191
    Side by side analysis if powered by coal 191
    Chapter summary 191

A Final Word.. 193

Appendix A: Electric Vehicles in Business... 194
    Types of electric vehicle available 194
    Is an electric vehicle suitable for my business? 196
    Charging facilities and commercial vehicles 198
    Electric vehicles and employees... 198
    Promoting your business with electric vehicles 198
    Chapter summary 199

Appendix B: Electric Vehicles and Local Government 200
    How local government can help the adoption of electric vehicles in their locality 200
    Do people use electric cars instead of bus services? 202
    Chapter summary 203

Appendix C: Other Electric Vehicles.. 204
    The future of transport? 204

  Other alternatives available today 207
  Chapter summary 209
Appendix D: Charging an Electric Car with Renewable Power 210
  Charging up an electric car with solar 210
  Solar-powered cars 211
  Solar power at home 212
  Commercially available solar charging-stations 212
  Finding out more about solar 213
  Wind power 213
  Chapter summary 215
Appendix E: Free – Working Towards a Radical Price 216
  Anyone want a free car? 216
  The mobile phone model 217
  Making the mobile phone sales model work for cars 217
Index 220

About This Book

This book is for anyone who is interested in owning an electric car and who wants to know more about them. If you want to know what they are like to own, use and live with on a day-to-day basis, I have written this book for you.

I've worked within the electric vehicle industry since 2004. I am an electric car owner myself and my family have been using electric cars since 2006. During that time, I've met and talked to hundreds of electric car owners and users. I've run one of the largest electric car owners' clubs in the world. I've worked with many car makers, advising them on their own electric car strategies, and I've been on the development teams creating new electric vehicles from scratch. This experience has given me a unique perspective of the industry, the cars and the owners who drive them every day.

Since the first edition of this book appeared in early 2010, the electric vehicle industry has changed beyond recognition. Gone are many of the quirky, low-speed electric cars like the G-Wiz; in have come all the mainstream car manufacturers with supremely good products. Many countries, including the United Kingdom, Israel, Portugal, Belgium, Estonia and Norway have nationwide high-speed charging infrastructures in place. Electric car start-up Tesla have gone from being a niche manufacturer of a sports car based on a Lotus Elise to a well-known premium car brand, their cars being compared positively with Aston Martins and Porsches by the motoring press and outselling the combined output of BMW, Mercedes, Audi and Jaguar in the luxury car segment. In short, the electric vehicle industry has gone from an inconsequential curio to a rapidly growing industry.

Now in its sixth edition, the book is the result of many years of research. It has been written with input from hundreds of people from all around the world. From Bangalore to Paris, from Los Angeles to London, electric car owners have contributed their opinions and their experiences about these exciting new vehicles. As well as existing owners, I have talked to vehicle manufacturers, car designers and electric vehicle infrastructure specialists from around the world to discover the technology that makes up an electric car.

I have spoken to members of the general public to understand their perceptions of electric cars. They have told me what they believe the positives and negatives of these cars to be, and what questions they would want answering before they would consider buying one for themselves.

I have worked hard to present a balanced picture of the environmental impact of electric cars. To do this I have worked my way through hundreds of research documents. I have met with electricity producers, key people in the nuclear industry, environmental pressure groups, car manufacturers and petrochemical companies to ascertain the benefits and detriments of the different technologies.

I have also carried out some real-world tests myself, comparing two electric cars with two comparable combustion engine cars to see how they measure up in terms of environmental performance.

While the environmental debate is an important one, it is not the sole reason for choosing an electric car. For many current owners, environmental considerations were not even a significant reason for buying their car.

This book is full of factual, relevant information, without the 'techno-babble' that all too often takes over the debate about electric cars. Where I feel it is relevant, I do talk about the technologies in an electric car, but only in order to help you make an informed decision about owning one.

The result is this book. By the time you have finished reading it you will understand what it is like to own, use and live with an electric car on a daily basis, and appreciate their benefits and drawbacks.

You will also know how to avoid the problems that some owners have experienced with their cars, ensuring that you get the best out of owning one.

Incidentally, just to avoid any confusion: although the book discusses hybrid and fuel cell cars, this book is specifically about battery electric cars: cars powered purely by an electric motor from energy stored in a battery and charged by plugging them into a power socket.

Foreword

by Robert Llewellyn

I have been intrigued, fascinated, confused, frustrated and amazed by electric vehicles for over 10 years. I've stared at battery meters as I drive through dark lanes, wondering if I'll make it home. I've woken up in the morning to find, as usual, that my 'tank is full' because the car has recharged overnight. I've driven over 10,000 miles in an electric car powered exclusively by solar panels. I've also charged my car from wind turbines and water mills as well as the national grid. It doesn't mind where the electricity comes from, it still works.

However I know tragically little about what really makes them go along the road and exactly how much energy they use per mile. Okay, I know they cost around 1p per mile to drive, but that's about it.

During the past five years I've met many people who've dedicated their professional lives to developing, improving and testing electric cars. I've tried to learn as much as I can about them, to try and gently re-balance some of the more absurd arguments against them, while understanding the perfectly legitimate anxiety and criticism they can create.

One man I have met in this process who helped me understand more about electric vehicles than anyone else is Michael Boxwell. He is not swept along by a wave of unfounded optimism, (an occasional failing of mine). He is a realist who knows from long, direct experience what a difference this new technology can make.

I have been driving electric cars for five years now. I've covered more than 50,000 miles in them and oh, and by the way, the batteries never left me stranded and weeping at the side of the road. However, Michael has been driving, developing and testing electric cars for far longer and is a wonderful source of genuine, trustworthy data.

If I'm ever really stumped on an issue of technological development, mechanical or electrical ignorance on my part, Michael is the man I run to.

Introducing the Electric Car

Just a year or so ago, spotting an electric car on the roads was something of a rarity. Today, it is a very different story. In the United States, there are over 400,000 electric vehicles on the road, with around 116,000 electric cars sold in 2015 alone. Across Europe, around 250,000 electric cars are now on the road and the numbers are growing every day. The figures are even higher in China, where over 300,000 small electric cars and vans were sold last year.

In some segments of the market, electric cars are best sellers: the Tesla Model S electric car outsells the BMW 7-series, Mercedes S-Class, Audi A8 and the Jaguar XJ-series *combined*. When the Tesla Model 3 was announced at the end of March, the company received orders for over 115,000 cars in the first few hours, and over quarter of a million people paid their $1,000 deposit in the first week. The days when owning an electric car was considered an oddity are very clearly behind us.

Electric car charging-stations are cropping up everywhere. Bollard-style charging posts and wall-mounted charging units have been installed in car parks and on streets in almost every major town or city, whilst the high-speed charging stations which can recharge your car in a few minutes are now installed on many major trunk roads and in service stations. Many countries, including the United Kingdom, the Netherlands, Japan and Estonia, have a country-wide network of high-speed charging-stations to allow you to drive across most of the country in an electric car. Most other developed countries are not far behind.

As prices fall, customers are responding and sales are now starting to grow rapidly. In the past year, new electric cars have been launched at ever lower prices, whilst the street prices of existing models, such as the Nissan LEAF, have dropped by as much as 30%. Combined with low-cost finance deals, this often makes purchasing an electric car as cheap as purchasing a conventional car, with a fraction of the running costs.

Electric cars are nothing new. The first electric car was built almost 175 years ago: predating the original combustion engine car by almost fifty years. At the beginning of the twentieth century, electric cars were popular. Owners liked them because they were smooth, quiet and easy to drive. They lost out to internal combustion engine cars because they were more expensive to buy and fuel became more readily available.

Back in the 1990s many manufacturers, like Ford, Toyota, Fiat, Honda, Citroen, Peugeot, Volkswagen and General Motors, all built tiny numbers of electric cars. Some were offered on sale or on lease to the public. Others were available on test to fleet customers. Some of these cars, such as the Toyota RAV4-EV and the Citroen Berlingo Electrique, are still on the roads today and occasionally find their way onto the used car market. By the early 2000s, however, all these cars had been phased out. Ten years ago, it was virtually impossible to buy a brand new electric car of any description, anywhere in the world.

Today, you can buy electric cars from Nissan, Ford, Mitsubishi, Renault, BMW, GM, Peugeot, Citroen, Smart, Chevrolet, Volkswagen and FIAT. Go to a busy modern Chinese or Japanese city and you will see electric vehicles everywhere you look. In Norway, electric cars already make up 5% of all the cars sold and regularly feature in the monthly best-selling lists. The market is growing and by 2020 many industry commentators believe that up to 10% of all cars on our roads will be powered by electricity.

It may still be early days, but the electric car is here to stay.

What is an electric car?

An electric car is a car powered by an electric motor, typically powered using energy stored in a battery. An electric car does not have a combustion engine. It is charged by plugging it into an electrical power socket.

There are electric cars of all shapes and sizes, from tiny city cars to premium luxury cars and SUVs. Some manufacturers sell electric powered variants of their existing cars, such as Volkswagen who offer electric versions of their Golf and Up! cars alongside their conventional models, whilst other manufacturers have built dedicated electric-only models, such as the Nissan LEAF, Renault Zoe or Chevrolet Bolt.

Electric cars do not have a gearbox. Instead, the electric motor powers the wheels, either directly or through a simple differential. As a consequence, electric cars have very few moving parts. Some electric car manufacturers claim that their cars have as few as 5-10 moving parts, compared to several hundred in a car powered by an internal combustion engine.

Electrical energy is stored in a bank of batteries built into the car. The batteries are relatively bulky and heavy, taking up about the same amount of space and weighing a little more than an engine and gearbox in a conventional car.

Charging time depends on the type of power socket you have. In Europe, most electric cars charge from a domestic socket in around 8-12 hours or from a dedicated home charging-point in 3-6 hours. In North America, where domestic sockets only provide 110 volts, a dedicated 240v charging circuit is installed in the home. Charging typically takes 3-4 hours for a high-capacity charger, or 6-8 hours for a standard charger.

Many electric cars also offer the option of rapid charging, to charge the car up in just a few minutes. Rapid charging works with dedicated high voltage, high current electricity charging units that are typically installed at car dealers or at service stations on major routes and can typically charge an electric car in around 25-30 minutes.

How an electric car compares to a conventional car

Whilst electric cars are not for everyone or every application, they are the ideal vehicle for many people and can offer significant benefits over other types of cars.

At the heart of an electric car is the electric motor. In terms of construction and power delivery, it is almost the complete opposite to a combustion engine:

	Electric Motor	Combustion Engine
Average efficiency	90% plus	25–35%
Maximum power	From standstill	At high speed
Multi-ratio gearbox required	Rarely	Always
Number of moving parts	2–3	130+

Because of these different characteristics, driving an electric car is slightly different to driving a conventional car. In the main, this means that electric cars are a more pleasant driving experience.

Advantages:

Maximum power from a standing start means an electric car pulls away smoothly and quickly. You do not have to rev the engine to pull away. There is no vibration from the motor and, of course, the electric motor is exceptionally quiet.

Because of the power delivery of the electric motor, most electric cars do not have a gearbox. This makes for an ultra-smooth power delivery and instant power whenever you need it. Throttle response is also much faster, typically five to ten times quicker than in a conventional car. Even in an electric car with fairly modest performance, such as the Nissan LEAF or VW e-Golf, this instant power makes for safer overtaking and gives the driver confidence that there is plenty of power in reserve.

When you take your foot off the accelerator, the motor helps slow down the car and recharges the batteries at the same time. The result is a much smoother and more progressive braking system, providing a more flowing transition between accelerating and braking.

An electric car is very easy to drive in heavy stop-start traffic. There is no gearbox or clutch to worry about and the car can crawl along at low speeds very efficiently with minimal effort on the part of the driver. Many electric car owners talk about how much less stressful an electric car is to drive in heavy traffic conditions. In the main, electric cars are in their element in built-up areas such as cities, urban and sub-urban environments, where speeds are limited and there is a lot of stop-start driving.

I have spoken to several people who suffer from travel sickness when travelling in a conventional car. They told me that they experience no problems in an electric car.

Disadvantages:

Some electric cars have a reduced top speed when compared to combustion engine cars. A few electric cars are designed purely for inner-city use and have performance to match. This is ideal to travel with the ebb and flow of traffic in a busy city, but not enough for longer and faster journeys.

When driving fast on high-speed roads and freeways, electric cars are at another disadvantage. The amount of energy required to travel at high speed means the range per charge can drop significantly compared to lower-speed use. Most electric cars can cruise quite happily at 70-80mph (110-130km/h), but the faster you drive the more quickly your range drops.

Range

You cannot discuss electric cars for very long before the discussion focuses on the range. All around the world, it is the number one concern that *non*-electric car owners have about owning one.

The reality is often very different from the perception. Many electric car owners will actually describe the freedom they feel every time they go out to their cars in the morning. They know they have enough range to go wherever they want to, without the hassle and cost of visiting the service station to refuel.

One electric car owner explains it like this. "It takes me nine seconds to charge up my electric car. That is the time it takes to plug the car in when I get home. The next time I need to use the car, it is charged up and ready to go." Several drivers talk about how convenient it is to plug their car in at home each night for a full charge the next day, as opposed to the inconvenience of driving to a service station to refuel every few days.

The majority of electric cars that are available today have a range of between 50 and 100 miles (80–160 km). This is far more than most people travel on a daily basis. According to the UK Department for Transport, the average car journey is 6½ miles (10½ km), with 93% of all car journeys being less than 25 miles (40 km). In the US, an average American driver travels 29 miles (46½ km) per day by car, with an average single journey of around 12 miles (19 km).

Many people never travel more than 50 miles (80 km) from their homes and many more will only travel further than 50 miles a few times each year.

For many new owners, the electric car will be bought to replace the second car in the family. In that case, range stops being an issue at all. If you need to travel further than the electric car can take you, just take the other car.

It is interesting to compare the concerns that non-electric car owners and existing electric car owners have about range:

- Non-electric car owners perceive that range is going to be a constant issue. They believe that they will be restricted because they cannot simply visit a service station to refuel their cars.

- Electric car owners like the fact that every time they go out to use their car it is fully charged up and ready to go. They have enough fuel to go wherever they need to and they will never have to visit a service station again.

Of course, there are people who regularly travel long distances to different locations in their cars and who do not have the luxury of a second vehicle. If you are an occasional long-distance driver, a network of charging-points, including high-speed rapid charging facilities, becomes important. If you are a regular long-distance driver, an alternative could be a plug-in hybrid electric car, which combines an electric drive with an engine to provide extra range for when you do have to travel further afield.

Charging it up

There have been various surveys carried out across the world on how and where people charge up their electric cars. Wherever the surveys have been carried out and whoever has done them, the one decisive figure is that around 95% of all electric car charging is carried out at home.

For most electric cars in Europe, the home charging-point is either a standard 220/240 volt domestic power socket or a dedicated home charging-station that can easily be installed at your home by a competent electrician. A full charge from a European domestic socket typically takes 6–10 hours, or 3–6 hours from a home charging-station, depending on which car you have.

Most car manufacturers dissuade you from plugging into a standard power socket to charge up your car, as the constant power demand for recharging a car from flat to full can be a strain on many household electrics. Indeed, a common problem with some early electric cars was that the domestic plug overheated and melted, fusing the plug and the socket together if left plugged in for several weeks! To resolve this problem, car manufacturers now restrict the charging speed if the car is plugged into a domestic socket. This resolves this particular problem, but does mean that regularly charging an electric car from a domestic power socket is very time consuming.

In North America, where the standard domestic socket only provides 110 volts, some electric cars can be trickle-charged from a standard socket, but most owners opt to have a 240 volt socket installed at their homes in order to charge up their cars in around 3–6 hours. Owners can still opt to trickle-charge their cars at 110 volts at their destination if they wish, as this can give them a useful range boost if they need it for the return journey.

Many electric cars can also be fast-charged using a dedicated rapid charging-point. These charging-points provide a much higher current in order to charge up batteries more quickly. A fast charge typically takes 25–35 minutes. These charging-points require a dedicated industrial power supply and therefore are unsuitable for home installation, but are being installed in major towns and cities, as well as at service stations on major roads.

Travelling longer distances by electric car

Compared to a few years ago, you are no longer limited to only using your electric car for shorter journeys. Thanks to a two-tier network of publically accessible charging-points, with high-speed rapid charging-points on major roads and a combination of high-speed and lower-speed charging-points in towns and cities, it is possible to top up your charge along your route or recharge your car once you have arrived at your destination.

Rapid charging allows you to recharge your car in around 25–35 minutes, whilst lower-speed charging allows you to fully charge your car up from flat to full in 3–6 hours, depending on the type of car you have.

Many countries, including the United Kingdom, Denmark, Norway, Estonia, the Netherlands and Israel already have a nationwide infrastructure of charging-points in place, whilst many states in the US are also getting close to having full state-wide coverage. Other countries lag a little behind but are catching up rapidly.

Charging at work

Many employees have made arrangements with their employers to charge their cars up whilst they are at work. Even if using a domestic power socket, it is possible to provide a good charge whilst at work for eight hours, effectively doubling the range of the car.

Occasionally, arranging a charging-point in this way can cause petty jealousies with work colleagues. Offering to pay for the electricity or donate money to charity usually resolves this problem. Alternatively, tell your colleagues to buy their own electric car so that they can save money as well!

Are electric cars the silent killer?

Every electric car owner I have met has said that they are always very aware of the lack of engine noise when driving. It is seen as one of the major benefits of electric cars. The lack of noise significantly reduces stress levels in many people, making the whole driving experience a lot more enjoyable.

When questioned, electric car owners say they are very aware that their cars are very quiet at low speeds. They say that they make sure they give other road users more space and take more care when driving in areas where there are a lot of cyclists and pedestrians.

When standing outside an electric car, it is not actually as quiet as you might expect. Some electric cars have a fan that runs whenever the motor is powering the car, which can clearly be heard by pedestrians. At anything over 10 mph (16 km/h), all electric cars make sufficient noise to be heard.

In built-up areas where an electric car is travelling so slowly that it cannot be heard, it is usually not the only silent vehicle on the roads. Bicycles are virtually silent as well, yet they are involved in fewer accidents with pedestrians than any other form of transport.

Furthermore, when driving at walking pace, many combustion engine cars are virtually silent and cannot be heard by pedestrians either.

The majority of the noise made by *any* moving car comes from road noise and wind noise. From outside the car, the noise of the engine is drowned out by these other sounds, unless the car is being raced.

Do you not believe me? Go and stand by the side of a road with high-speed traffic. What traffic sounds do you hear?

- The 'whoosh' sound is the sound of the air splitting and then reforming around the car as it travels.

- The lower-pitched rumble, similar to the friction sound when you drag your feet slowly along the ground, is the sound of the rubber on the road surface.

- Unless a conventional car is in a low gear and accelerating heavily, the chances are you will not hear the engine until the car is next to you, and possibly not even then.

The claim that electric cars are 'silent killers' appears to be a myth. Concern about this being a future problem appears to be a perceived issue resulting from poor quality research rather than a real problem. I have not been able to find records of any deaths or serious injuries anywhere in the world that have been attributed to the silence of an electric car.

Back in 2010, I asked a member of the UK Parliamentary Advisory Council for Transport Safety (PACTS) for his own opinion about the issue of silent electric cars, and his personal belief was that this is not an issue.

I also spoke with a car designer who used to work with blind people. His personal view is that electric cars pose no greater risk to blind people than any other road transport.

Since then, I have been involved with two research projects that have specifically looked into the noise issue with electric cars. Both projects have concluded that, even at walking pace, electric cars pose no more of a threat than conventionally powered cars travelling at the same speed.

Interestingly, both projects also concluded that conventionally powered cars are also extremely quiet when driven at walking pace. If a supplementary noise is required for electric cars, it may be that conventional cars will also require this noise in order to remain safe.

One of the projects included a drive-by sound test comparing an electric Nissan LEAF with a similarly-sized Citroen C4 equipped with a conventional internal combustion engine. Both cars were driven down a road at 20 mph, 30 mph and 40 mph, and the noise levels recorded by the roadside. To our astonishment, the Citroen C4 was marginally quieter than the Nissan LEAF at all three speeds, due to lower levels of road noise produced by the Citroen.

Less scientifically, I have been using electric vehicles for eight years and have driven an electric car on a daily basis for the past six. I have not once witnessed anyone step out in front of me unexpectedly, nor experienced a problem caused by pedestrians not hearing me.

However, it is also true to say that this is not a subject that it is worth being 'dead right' about. The United States Government has decided to act now rather than risk a potential problem in the future, when there are a lot more electric cars on the road. The *Pedestrian Safety Enhancement Act* became law in the United States in January 2011, to ensure that new electric vehicles provide an audible warning to alert pedestrians that there is a vehicle in motion.

Consequently, every electric car on sale in the United States now has a sound system in place to alert pedestrians. Most manufacturers fit all their electric cars with this system, even for cars sold outside of the United States. The Nissan LEAF sound system is typical in its operation. When travelling at walking pace, a sound system plays an engine note that can be clearly heard from outside the car, yet is virtually undetectable from within the car itself. As the car accelerates, the electronic sound fades away as the wind and road noise become more apparent.

Other forms of electric cars

If you regularly need to travel long distances and range is an issue, a hybrid or range-extended electric car may be the solution. Hydrogen-powered cars may be suitable in the longer term, but at present these are only available in parts of California and Iceland.

Hybrid cars

Hybrid cars combine an electric motor with a combustion engine, using the electric motor to assist the combustion engine during acceleration and for low-speed driving.

There are three alternative types of hybrid cars:

Parallel hybrids

Hybrid cars like the Toyota Prius, the Mitsubishi Outlander PHEV and all the Lexus hybrids are known as *parallel* hybrids. The electric motor and the engine work in parallel to power the car. In other words, either the electric motor or the engine can turn the wheels.

Series hybrids

A *series* hybrid car is different. A series hybrid is an electric car where only the electric motor powers the wheels. The combustion engine then powers a generator to run alongside the batteries when required. This technology has been in use in diesel–electric trains for the past fifty years.

Series hybrids are the basis for the Chevrolet Volt (North America) and the Opel/Vauxhall Ampera (Europe). The BMW i3 hybrid is a series hybrid too: the car runs as a fully electric car and the engine kicks in to keep the batteries charged when a longer range is required.

Twin drivetrains

Twin drivetrain hybrids are vehicles where two wheels are powered by a conventional internal combustion engine and gearbox, and the other two wheels are powered by an electric motor.

The electric drive and the conventional drive are kept entirely separate. The cars can either run on electric power, by using the internal combustion engine, or with a combination of the two.

The Volvo V60 hybrid and the Peugeot and Citroen hybrid cars that are now on sale in Europe are all based on this technology.

Range-extended electric cars

The Chevrolet Volt and its European twin, the Opel/Vauxhall Ampera, were the first of the range-extended electric cars. Launched in 2010, they are electric cars that can be plugged in to recharge, with an electric-only range of up to 40 miles (65 km). Once the batteries are flat, a combustion engine kicks in to recharge the batteries.

Volvo launched the V60 Hybrid in Europe in 2013, and their XC60 SUV Hybrid arrived the following year. Unlike the Chevrolet Volt, the V60 uses a diesel engine in combination with an electric motor, providing superb fuel economy. In eco-hybrid mode, Volvo claims economy figures of up to 150 miles per gallon[1]. I have driven the V60 Hybrid in Sweden and came away very impressed. In electric-only mode, it has a range of up to 30 miles (45 km). In performance mode, it can sprint from 0 to 62 mph (0 to 100 km/h) in 6.9 seconds.

The Mitsubishi Outlander PHEV gives a range of up to 32½ miles on electric power before switching over to its internal combustion engine.

Toyota, Ford, BMW, Honda, Mercedes, Mitsubishi and Porsche also have range-extended electric cars. The Toyota Prius Plug-In Hybrid has a range of up to 14 miles (22 km) in electric-only mode, whilst US customers can also purchase a plug-in hybrid versions of the Ford Focus or Honda Accord with electric-only ranges of up to 13 miles (21 km) and 21 miles (34 km) respectively. The Porsche Panamera S E-Hybrid, with

---

[1] The Volvo V60 Hybrid 150 mpg figures are based on the imperial measure of 4.546 litres to the gallon. The US gallon contains only 3.8 litres. Based on the US gallon, the Volvo V60 Hybrid provides economy figures of up to 125 mpg.

an electric-only range of 20 miles (32 km) combines its high-performance V6 supercharged engine with a 95 bhp electric motor to provide a combined 91 miles per gallon[2]. The new Mitsubishi Outlander PHEV is a seven-seat SUV with an all-electric range of up to 32½ miles (52 km) and a combined 148 miles per gallon[3].

Although the efficiency of these vehicles is not a match for full electric vehicles, plug-in hybrids have the potential to offer a good solution for people who want an electric car but for whom the ability to travel longer distances on a regular basis is also important.

Do hybrid cars provide the best of both worlds?

Hybrid car enthusiasts claim that hybrids provide the best of both worlds, combining a practical long-distance car that can cruise at freeway speeds all day long with an electric car for town use. This is undoubtedly true with the plug-in hybrid cars. If you have to travel long distances, you have a real benefit in being able to switch over to electric power for relatively long periods of time. For non-plug in hybrids such as the Toyota Prius, this benefit is significantly reduced, as they only have a very short electric-only range – typically 1–2 miles.

Many hybrids (sometimes referred to as *mild hybrids*) use the combustion engine all the time, using the electric motor only to supplement the engine. These cars cannot be used in 'electric car only' mode.

With most hybrid cars, the batteries are charged up using both the combustion engine and regenerative braking, although many of the plug-in hybrids only fully recharge their batteries when they are plugged in. Although at lower speeds using an engine to

---

[2] The Porsche figures of 91 mpg is based on the Imperial gallon of 4.546 litres to the gallon. The US gallon contains only 3.8 litres. Based on the US gallon, the Porsche provides economy figures of up to 76 mpg.

[3] 148 mpg based on the imperial gallon. Based on the US gallon, the Mitsubishi provides economy figures of up to 123 mpg.

charge batteries is more efficient than using an engine to drive the wheels, this means that the electric element of the vehicle is still powered using fossil fuels.

The production cost and energy required in producing hybrid cars is higher than for other cars, as they have an internal combustion engine and an electric motor plus batteries in the same vehicle. However, this is still a relatively insignificant impact on the environment when compared to the overall running costs of a car. I cover the environmental impact of car production in a lot more detail later on (starting from page 179).

For regular long-distance driving, where pure electric cars are currently not suitable, hybrid cars, especially the plug-in range-extended hybrids, do have several benefits over conventional combustion engine cars.

How hybrid cars compare with economical diesel cars

Across Europe, diesel-powered cars have become popular for long-distance driving. A modern diesel car is very economical and has comparatively low carbon dioxide emissions. From a driving perspective, they can be ideal long-distance vehicles.

From an economy perspective, diesel-powered cars can provide similar long-distance economy to current hybrid cars. For this reason, a lot of people believe that hybrid cars are irrelevant.

However, this only shows part of the benefit of a hybrid over a diesel-powered car. For long-distance driving, hybrid cars and diesel cars provide similar economy. Around town, a hybrid car produces significantly improved economy, as the electric motor is far more efficient at low speed and stop-start driving than a diesel engine.

The biggest benefit of hybrid cars over diesel cars comes from the improved overall emissions from the car, of which carbon dioxide is just one. The health impacts of diesel pollution are now well understood, particularly in built-up areas where diesel pollution has been identified as the primary cause of tens of thousands of premature deaths each year. We will cover the issue of diesel pollution in more detail later on in the book, when we look at Electric Cars and the Environment (see page 131 for more information on diesel pollution).

Hydrogen fuel cell cars

The Toyota Mirai is a new hydrogen fuel cell car being launched into selected markets in 2016.

Hydrogen fuel cell cars are electric cars where hydrogen-powered fuel cells generate electricity to top up the batteries. The driving characteristics of fuel cell vehicles are the same as for any other electrically powered vehicle.

A fuel cell is a power generator that produces electricity through a chemical reaction with the fuel (i.e. hydrogen), as opposed to burning the fuel as with a combustion engine. Fuel cells typically extract twice as much energy from hydrogen than burning hydrogen in an engine. However, the chemical reaction takes much longer than burning, and therefore fuel cells create energy at a much slower rate.

At present, very few hydrogen fuel cell cars exist. The Honda FCX Clarity has been trialled for a few years in California, the Mercedes-Benz 'F-Cell' cars were trialled in Berlin until the end of last year. The new Toyota Mirai is being launched in selected markets in 2016. The perceived benefit of fuel cell cars over electric cars is that they can be refilled quickly at any available hydrogen fuelling station. In reality, hydrogen fuel stations simply do not exist in any great number. Worldwide, there are currently 177 hydrogen service stations open to the public, only 55 more than there were at the beginning of 2012. The majority of these are in Iceland, with a handful more in California, Germany, Japan and the United Kingdom.

Many experts now believe that hydrogen is unlikely to become common for cars over the short or medium term, and that hydrogen fuel cells are most likely to be superseded by more advanced batteries and faster battery recharging facilities over the next five years.

There are also question marks about the environmental benefits of hydrogen-powered cars:

- Hydrogen fuel cell cars are emission-free at the point of use, but the extraction of hydrogen in the first place is a very energy-intensive process.

- For instance, extracting hydrogen from water through electrolysis requires three units of electricity for every unit of hydrogen you generate.

- This is not a problem in countries where electricity is abundant and carbon friendly. Iceland, for example, has already installed hydrogen fuel pumps at service stations across the entire country.

Using hydrogen in a combustion engine

Several manufacturers, including Ford, Mercedes and BMW, have developed prototype cars and vans that can use liquid hydrogen in a modified combustion engine car. These promise zero emissions from the vehicle, using a combustion engine rather than fuel cells.

It is also possible to convert an internal combustion engine car to run on hydrogen. This is becoming popular in Iceland, where the government has already invested in hydrogen service stations. The hydrogen is generated using environmentally friendly geo-thermal energy.

The cost of building hydrogen-powered combustion engine cars is much lower than building dedicated fuel cell cars. The cars are built using existing mass-produced technology.

However, hydrogen-powered combustion engines are less than half as efficient as hydrogen fuel cells, which means you have to stop to refuel your car every 100–150 miles.

Chapter summary

- Electric cars are increasingly regarded as being the *next big thing* by the car industry and governments alike.

- Electric cars are available today from a wide choice of manufacturers.

- Electric cars have different driving characteristics to conventional cars.

- Electric cars are in their element in cities and urban areas, although that does not stop them being driven elsewhere.

- Non-owners have concerns about range. This is rarely an issue with existing owners.

- There are various options available for travelling longer distances.

- Hybrid cars reduce the environmental impact of the combustion engine.

- Hybrids are significantly better than diesel-powered cars in terms of environmental efficiency, and in particular have some significant benefits over diesel-powered cars in towns. They currently offer the best solution for frequent long-distance drivers, from an environmental viewpoint.

- Hydrogen fuel cell cars are electric cars with a fuel cell generating power that tops up the batteries, which in turn power the car.

Why are Electric Cars Important Today?

The possibility of an electric vehicle was first demonstrated in 1828, when a Hungarian scientist, Anyos Jedlik, built a tiny model car powered by an electric motor that he had invented. In 1835, Dutch professor Sibrandus Stratingh built a small electric car with non-rechargeable batteries. By 1881, electric tricycles were on sale in Paris. By the turn of the last century, electric cars were becoming established and gaining in popularity.

Early electric cars were popular because they were quiet, clean and easy to operate. But they were also expensive, slow and heavy, and as the internal combustion engine became more powerful, reliable and cheap to produce, electric vehicles very quickly faded away, becoming an obscure footnote in the history of the motor car.

So what has changed? Why are electric cars relevant now, when they have been largely ignored for the past century? Are they simply a political tool to appease climate change activists, or are they a practical form of transport that can be justified on their own grounds?

The changing face of transport

There are lots of reasons why electric cars are relevant today. Inner city pollution, the cost of fuel and traffic congestion all play their part, but probably most importantly of all, there is a subtle shift in perception of the role of the car in society today.

Owning and running a car *used* to be relatively cheap. Wind the clock back twenty years and fuel was cheap, car insurance was cheap and, even if you were a young driver with a modest amount of money to spend, you could afford to own and run a car. I remember as a student I could afford to buy and run a cheap second hand car, earning money with a Saturday job. Most of my student friends were the same.

Today that is no longer the case. Cars are expensive to buy and run. Students ride bikes or take public transport. Many people on low incomes simply cannot afford to run a car. Many car owners have downsized their cars; instead of buying family cars with big, thirsty engines, they buy smaller, more economical models.

Means of personal transportation such as bicycles, electric bikes, scooters and mobility vehicles have all seen huge sales increases over the last few years. Motor manufacturers are experimenting with new types of vehicles, such as the Renault Twizy, a small lightweight vehicle that is somewhere between a motorcycle and a car. It all points to the possibility that the car of tomorrow could be a very different vehicle to the car of today:

Energy security and price volatility

Oil prices have crashed over the past year with the cost of a barrel of crude oil down to well under $30 a barrel. Prices currently swing between $25 and $35 a barrel, down from over $100 a barrel in June 2014.

Whilst this is good for the consumer at the fuel pumps, the true story about what is happening within the oil industry is worrying news for everyone.

Five years ago, America was a net importer of oil. By 2013, America was producing as much oil as it used, thanks to shale oil production. The International Energy Agency in 2013 suggested that shale oil would help global oil production grow by around 8.4 million barrels a day, outstripping demand and making America the number one oil producer by 2020, becoming a net exporter of oil by 2030[4].

These were alarming figures for countries in the Middle East, who could see their long term profits and income under threat. Oil prices dropped thanks to over production, which was then exasperated by an OPEC-lead price war. Traditional oil production is comparatively cheap, whilst shale oil and underwater oil extraction is significantly more expensive. By forcing the price of crude oil down to under $40 a barrel, both shale oil in the United States and North Sea oil in the United Kingdom cannot be produced economically. By keeping the price down over an extended period of time,

---

[4] International Energy Agency World Energy Outlook 2012.

it is believed within the industry that OPEC hope to force out other oil producers. If this is indeed their strategy, it is certainly working. Investment into shale oil production has significantly reduced, whilst investment in North Sea oil production in the United Kingdom has virtually dried up.

It certainly would not be the first time that the United Kingdom or the United States has been held to ransom over oil. And whilst nobody is predicting another oil supply crisis like we witnessed in the 1970s, there is absolutely no doubt that when the oil barons decide that cheap oil must come to an end, the West is going to have to pay.

Peak oil

'Peak oil' is the point in time when the maximum rate of petroleum extraction is reached, after which the rate of production enters a terminal decline. It was originally identified in 1956 by M. King Hubbert, a geo-scientist working at the Shell Research Laboratory in Houston, Texas.

M. King Hubbert calculated that peak oil in the United States would be reached in 1965 and that worldwide peak oil would be reached in 1995 "if current trends continue". Originally ridiculed for his research, he became famous in the mid-1970s, when his predictions about US peak oil were proved correct.

M. King Hubbert's predictions about worldwide peak oil being reached in 1995 were not so accurate. New oil fields were discovered in the intervening years and oil companies learnt how to extract far more oil from each field than was previously possible.

The entire peak oil theory came in for criticism in 2013, as shale oil production built up, particularly in North America. Worldwide oil production is still on the increase, thanks to fracking and shale oil from North America and modest increases in Russia, China, Pakistan and Chile[5]. In 2013, worldwide production was estimated at 75.3

---

[5] Oil and Gas Journal: Worldwide reserves, oil production post modest rise: 12/02/2014.

million barrels of crude oil per day, up from 74.7 million barrels per day in 2012 and 73.6 million barrels per day in 2011.

Yet shale oil and improved production techniques do not make significant changes to the peak oil theory. When the International Energy Agency revisited their figures for their 2013 report, they identified serious flaws with their own calculations in the previous year's report. Their earlier report took into account neither the decrease of production from other oil-producing nations nor the significantly higher cost of extracting oil from new sources, which makes them uneconomical.

According to the Oil and Gas Journal reports from February 2014, thirty-seven countries are now past peak oil, including virtually all major oil producers other than Canada and the United States[6].

Meanwhile, worldwide demand for oil continues to rise, and is rising faster than production. This has increased rapidly with the falling price of oil over the past year. Whilst at the moment, the oil industry is able to supply all of this increased demand at the present, it is widely believed within the oil industry that the current levels of production are unsustainable for more than three to five years. When worldwide peak oil is reached in the next few years, the high demand, combined with *falling* supply, will rapidly see prices rising to significantly higher levels.

All this makes finding an alternative fuel for transport both more important and more viable. With crude oil prices increasing so significantly, the cost of buying and running an electric car become more appealing. In Europe, we have already reached the point where the entire purchase and running costs for their electric car are, for many people, comparable to the cost of paying for fuel in a conventional car. As fuel prices continue to rise, more and more people will find themselves financially far better off running an electric car.

---

[6] Oil and Gas Journal: Worldwide reserves, oil production post modest rise: 12/02/2014.

The multi-car family

Many households have two, three, four or even five cars. Each car is used by a different family member. One of the cars will be nominally the 'family car', which is used when the whole family go out together. The other cars will be used for day-to-day journeys, most of which are repetitive and usually quite short.

Because the journeys of these 'second cars' are often repetitive, it is easy to work out exactly how far each car travels per day: it's usually somewhere between 15 and 22 miles (24–35 km) each day: an ideal distance for an electric car. As the price of electric cars continues to drop, more and more families are starting to replace their second cars with electric cars and pocketing the savings in running costs.

The generation gap

Twenty years ago, it was every young person's dream to learn to drive and own their first car. Today, for many young people that is no longer the case. Partially, this is due to the costs of car ownership: fuel, insurance and tax all cost much more than they used to, but there is a bigger underlying cultural change that is having a greater effect.

People's concept of freedom is changing. In the past, a young person's first car was their first step to freedom. Everyone from my generation can remember their first car and how they used it to meet up with their friends and do what they wanted.

Today, young people do not associate freedom with travelling to see their friends. Instead, their friends are on Facebook, Skype or WhatsApp. They can play games with them across the internet via their mobile phones. When they meet together, they're more likely to play computer games than they are to go out for a drive.

Video games have also changed the perception of driving. High-speed car racing games give people the opportunity to drive Ferraris and Lamborghinis at break-neck speed through Monte Carlo, from the comfort of their sofas. Research shows that you get the same feelings of excitement driving a race car on a computer game as you would if you were driving that same car for real[7]. In comparison, the excitement of driving a cheap, rusty old car for real isn't quite as alluring a proposition as it used to be!

The concept of ownership is different as well. Young people have grown up with free stuff: they get a brand new mobile phone for free every two years; they listen to music on Spotify for free and they watch videos and films on demand, also for free, on YouTube. They get their e-mail for free from Google or Hotmail. What they do pay for they share with their friends: video games, magazines, books, even clothes get swapped around. It has changed the way younger people view their possessions. Research shows that when they do start driving, young people are far more likely to share their car with their friends. They are also the most likely age group to sign up to a car-sharing club rather than owning their own car.

So technology has changed our culture. For young people of every generation, freedom means the chance to spend time with friends. The way we do it has changed: a car is no longer an essential part of the equation, and certainly not as important as an iPhone or a reliable internet connection.

Consequently, what many young people want from a car has changed as well. A car has got to look good, it's got to be fun and interesting, but it essentially it is just a tool to do a job. Consequently, there are fewer modified cars on the road, kit cars have almost become a thing of the past and young people are more interested in cost of ownership than performance figures.

The way young people view the car also changes the ownership model. They've grown up with the concept of 'pay as you go': they get a free phone and just pay to use it. Consequently, they want their cars to be the same. Yet that model doesn't work well for conventional cars. You have to pay for the car, pay to service and maintain it, pay to fuel it, pay to insure it. An electric car, however, can work a different way. With fuel costs removed from the equation, an electric car can be sold in exactly the same way

---

[7] Which World is Real? The future of virtual reality – *Science Clarified*: 2014

as a mobile phone: choose how far you wish to travel each month and get the car, the maintenance and the insurance for one fixed monthly cost.

Pollution

Whilst carbon emissions from road transport have fallen dramatically over the past twenty years, inner city pollution remains high in many cities across the world. And whilst carbon levels have dropped, the switch from petrol to diesel for many cars, particularly in Europe and parts of Asia, has significantly increased nitrous and particulate matter pollution, including many substances that are harmful to human health and are estimated to cause around 28,000 premature deaths per year in the United Kingdom alone[8] (see page 131 for more information on diesel pollution).

Electric cars move the pollution out of the cities and towards the power stations creating the power. The way your electricity is produced will determine how much less pollution an electric car is responsible for compared to a conventional car, but the fact that fewer pollutants are being emitted inside the city is a significant step forward in public health.

The rise of alternative electric vehicles

It's a mistake to think that the vehicle that replaces the conventional car will be a car as we know it today. Driving an electric car doesn't stop you getting stuck in a traffic jam. It does not reduce the amount of traffic on our roads. If we want to make a real difference in our cities, a different type of vehicle is required.

These different vehicles already exist in various shapes and forms:

---

[8] Source: Estimates of Mortality in Local Authority Areas Associated with Air Pollution. Published by Public Health England, April 10th 2014.

Electric Bicycles

All around the world, people are switching on to electrically assisted pedal cycles. Around 325,000 were sold in the United States last year, up from 250,000 in 2014. In Europe, the figure was over 800,000. Research suggests that the average electric bike owner uses their bike at least five days a week and that the average journey is around two miles. In seven out of ten cases, the electric bike was used instead of travelling by car[9].

I have been involved with the development of an electric pedal cycle for inner city deliveries. Called the Cargo Trike, it has a 250 kg (550 pounds) payload and can legally use cycle lanes. During testing with a courier company in London, the customer found they could make up to 160 deliveries a day with the Cargo Trike, compared to an average of 50-60 deliveries with a conventional small van, simply because the Cargo Trike was easier to use, could cut through traffic and was more convenient to park.

Mobility scooters

Overlooked by many, the humble mobility scooter has become a huge hit for less able-bodied people. These can legally be used on footpaths and in shops and many of them can also be used on the road. They provide freedom for a lot of people who would otherwise not be able to get out and about.

Worldwide, around ten million mobility scooters are sold each year. There are around 300,000 in the UK alone, with a similar number in the United States. Owners use their vehicles most days and often prefer using them to driving a car, finding the convenience of a mobility scooter more suitable for short journeys.

At present, most countries restrict the use of mobility scooters to people with disabilities, but new, more stylish vehicles are in the pipeline that are likely to increase the potential interest in these vehicles. Combined with autonomous driving technology

---

[9] Source: Bike Europe market report, 5th May 2014.

(see below), it is not inconceivable that a future general purpose 'personal mobility vehicle' could become a popular solution for short distance urban journeys.

Autonomous (self-driving) vehicles

Self-driving vehicles are gathering a lot of media attention at the moment, with the Google self-driving cars clocking up significant mileages in California, Texas and Washington DC. Pilot schemes are taking place in Coventry and London in the United Kingdom and various other tests are being carried out by most of the major car manufacturers.

Yet it may come as a surprise to many people to know that the technology is already in widespread use for off-road applications. Autonomous vehicles exist in factories and on large campuses for moving components around the site; self-driving shuttle buses or trains are a common sight at some airports and many farm tractors have self-drive modes for ploughing, sowing, inspecting and harvesting crops.

Much of the technology for self-driving vehicles already exists in mainstream cars. The Toyota Prius has had a self-parking mode for years. Volvo have incorporated self-driving technology into many of their latest cars, using it to warn the driver if there is the potential for an accident, then taking over the control of the car by braking and steering away from the collision if the driver does not take action themselves.

Fully self-driving cars are only a few years away. The trials are already advanced and car manufacturers are now talking about having self-driving cars on sale within the next few years. Volvo has pledged to have a self-driving car on sale in Europe by 2017. In the UK, one manufacturer is very advanced in their design for an electric single seat 'personal vehicle' that is entirely self-driven.

Self-driving vehicles have the potential to radically change the way we think about cars, particularly in a city environment. Autonomous taxis could be parked on every street, in every city. When you want to go somewhere, you could summon one using your mobile phone and it would be parked outside your house ready for you by the time you'd walked out of your front door. Once it has taken you to your destination, it drives off for its next customer. Self-driving vehicles could be used to collect children from school, take elderly people to the shops and carry out all the usual, everyday journeys, all at the fraction of the price of owning a car outright.

Shared self-driving cars solve many of the problems of congestion. With widespread take-up, significantly fewer vehicles are required in the first place, needing fewer parking spaces. Self-driving vehicles can communicate with each other to avoid congestion, always taking the shortest or quickest route to your destination, reducing the blocking points in a city and keeping traffic flowing effectively.

The potential for self-driving vehicles is potentially limitless. Some experts are predicting that such vehicles will make the current car ownership model obsolete within a decade.

Chapter summary

- Electric vehicles of all types have an important and relevant part to play for the future, not just at governmental and policy level, but at an individual level as well.
- The change of attitude towards the car, the ever-increasing costs of running a car and the availability of other options, particularly personal mobility vehicles, are already changing our driving and purchasing habits.
- It is unlikely that electric vehicles will ever completely dominate the car market, nor would that necessarily be a good thing if they did, but they are already of importance and value, and that importance and value will only increase as time goes on.
- Don't expect the car of the future to be the same as the car of today. The uptake of personal mobility vehicles (electric bicycles and mobility scooters) and autonomous driving have the potential to create a huge shake-up in personal transportation.

Living with an Electric Car

First impressions

There is a well-known phrase within the electric vehicle industry: *The EV Grin.* It refers to the involuntary smile that everyone has when they take their first ride in an electric vehicle. It is a very strange sensation, getting into an electric car for the first time and driving away. The silence, the smoothness and the lack of vibration take some getting used to. Everyone's first reaction is '*Wow!*'

Most electric cars do not have a gearbox, so the gear lever usually resembles an automatic, allowing you to select forward, reverse and park. In some cases, the gear lever has been replaced by a simple switch for selecting forward or reverse.

Once you switch on the ignition, there is no noise or vibration to let you know the engine is running. Most people are unsure whether the car is actually ready to pull away. As you tentatively put the car into *Drive* and touch the accelerator, there is a mild sense of shock that the car pulls almost silently away.

It takes a few minutes to get used to the strange sensation of travelling without engine noise. At speeds below 10 mph (16 km/h), the car makes very little noise, and as a driver you are aware of that. In fact, you will probably be more aware of that than anyone around you, as most modern cars are very quiet at walking speeds, anyway.

As the speeds increase, road noise and wind noise now mean that your electric car is making virtually the same amount of noise to the outside world as any other car. Inside the car, the impression is still that the car is exceptionally quiet, because you cannot hear an engine and modern cars have high-quality insulation to keep road and wind noise out of the cabin as much as possible.

Performance

Once you are used to the silence, you start noticing other things. The power delivery is very smooth and there is no vibration from the engine.

As you drive the car, you will start noticing how good the low-speed performance is. Electric cars are usually very quick when pulling away, and driving around town can be an enjoyable experience. The vast majority of electric cars provide excellent around-town performance.

At higher speeds on open roads, some of the smaller electric cars may feel underpowered. This is not the case with all models, but often high-speed performance is adequate rather than rapid.

The fun factor

Most people who have never driven an electric car are quite surprised by how much fun they can be.

A lot of the latest generation of electric cars are remarkably quick. The Nissan LEAF has as much torque as a conventional car with a 2½ litre V6 engine, while the Mitsubishi i-MiEV and Peugeot iOn have 0–62 mph (0–100 km/h) acceleration figures that are at least one second faster than the equivalent internal combustion engine cars. The BMW i3 is even more rapid, with sub 8-second 0–62 mph acceleration, whilst the Tesla electric performance cars are amongst the fastest cars on the road today.

This performance is always instantly available. Unlike a car with a gearbox (even an automatic), the car is always capable of providing instant acceleration whenever it is needed. There is no delay after putting your foot on the accelerator as the fuel is pumped into the engine. There is no changing gear to make sure you have enough power. With an electric car, the power is available instantly.

As the heavy batteries are mounted low down in the chassis, most electric cars have a low centre of gravity, often making for reasonable handling. Cars like the BMW i3, the Renault Zoe and Mitsubishi i-MiEV are great fun to drive. The instant acceleration, responsive steering and sharp handling make for an entertaining and enjoyable driving experience.

Braking

Most electric cars have two braking systems, combining *regenerative braking* with a standard braking system. Regenerative braking uses the momentum of the car to generate power to put back into the batteries.

Different manufacturers implement regenerative braking in different ways. The best systems build regenerative braking into the brake pedal of the car. To slow down, you put your foot on the brake pedal and all the power is then fed back into the batteries to extend your range. The mechanical brakes are only deployed if you have to stop quickly.

In other electric cars, regenerative braking kicks in as soon as you take your foot off the accelerator pedal. In cars like the Tesla Roadster or the BMW i3, you can actually come to a complete standstill simply by taking your foot off the accelerator. It takes a little getting used to, but it does make for efficient driving.

Some cars allow you to adjust the level of regenerative braking, so that the energy that would otherwise be lost in braking is used to recharge the batteries instead. In the Mitsubishi i-MiEV and the Nissan LEAF, you can adjust the level of braking using the gear shift to provide little or no regenerative braking or to increase the braking levels so that most of the stopping power comes from the regenerative system.

Range fixation

You will not be far into your first drive before you become aware of the fuel gauge. In fact, if you have never driven an electric car before, the fuel gauge becomes a fixation for the first few weeks.

An electric car has a shorter range than a conventional car with a fuel tank, so the fuel gauge moves more quickly than you will first expect.

In most conventional cars the fuel gauge seems to stay close to full for the first half tank of fuel and then moves down quite rapidly afterwards. In any car, once the fuel gauge drops below a quarter, many people start getting 'range fixation'. They are on the lookout for a fuel station and start to worry if they cannot find one on their route.

In an electric car, the fuel gauge will start to move after only a few miles of driving. It is a bit disconcerting at first, because everyone is so used to the way fuel gauges work in conventional cars. In effect, you are getting the same range fixation as you do when running low on fuel in any other car.

Even when you are driving a short distance and you absolutely *know* there is enough charge to get to your destination, it is very easy to get fixated on range in the early days.

It is a psychological difference. In reality, you are no more likely to run out of range in an electric car than you are to run out of fuel in a normal car. The best analogy to use is that of a mobile phone. If you plug your phone in overnight to charge it up, it has enough charge to last the next day. The same is true with an electric car. Once you have more confidence in your electric car and have used it for a while, your range fixation disappears.

In fact, you get to the point where you ignore the fuel gauge completely. After all, if you plug your car in every night and you know that you have enough range to do your daily driving, why bother checking the fuel gauge?

Plugging it in

Most electric cars have a socket on the outside of the car. Sometimes this is hidden behind a flap in the nose of the car; sometimes it is behind a fuel filler cap.

In Europe, some electric car owners use a normal domestic household power socket to charge up their cars. The standard domestic power in Europe is 220/240 volts, which is enough to charge up most electric cars in 8–10 hours.

In North America, where domestic household power is only 110 volts, you can charge up an electric car from a domestic power socket, but a full charge is likely to take between 16 and 24 hours for most electric cars, or around 6–8 hours for a *Neighborhood Electric Vehicle* (NEV). Sometimes 110 volt charging is known as 'Level 1 charging'.

In order to speed up this charge time, a higher voltage is required. Thankfully, most American households have a 240 volt feed, which means that an electrician can install a higher-voltage socket so that you can charge up an electric car in 7–8 hours, or an NEV in 4–5 hours. Charging from a 240 volt supply is often known as 'Level 2 charging'.

In both Europe and North America, more powerful *Level 2* electric car chargers are also available to speed up the charging process further. These still work at 240 volts, but run at a higher current. Depending on the capacity of the charger and which car you have, these home charge units will allow you to recharge your car from flat in around 3–6 hours.

You can install home charging units either inside or on an external wall of your home. Likewise, if you are just using a domestic power socket, you can buy an external socket to mount on the outside wall of your home for convenience.

Very rapid charging systems are also available to charge up an electric car. These are extremely high-voltage, high-current systems, designed to provide a boost charge in around 25–35 minutes, and are sometimes known as 'Level 3 chargers'. The cost and complexity of these units means that it is not practical to have one of these charging systems at your home, but you will see them at some public charging-points. These Level 3 chargers are the backbone of a public charging infrastructure and many countries, including the United Kingdom, now have a complete network of high-speed charging-points, installed at motorway and dual carriageway service stations, in cities, at selected car dealerships and at some petrol stations.

It feels very strange to plug in your car for the very first time. It is a novelty that takes some time to wear off, but the actual process of charging up a car is no more complicated than charging up a mobile phone.

The first few weeks

The first few weeks with an electric car are fun. The novelty factor of a car that runs on electricity lasts a while and, in general, most electric cars are a lot of fun to drive.

You go through a period where you are thinking about every journey you are taking. You are constantly checking that you have enough range and often making arrangements to plug your car into a power socket at your destination, even if your destination is well within the range of the car. It is all part of the *range fixation* that you get as a new electric car driver. It soon wears off.

One thing that happens to most people at some point in the first few weeks is that they forget to plug the car in. Most people do it once and usually at the most inconvenient time. Thankfully, if your journey is relatively short, you can often plug your car in and get enough range after 20-30 minutes of charging to get you to your destination. As a consequence, the results are rarely catastrophic – but most electric car owners rarely make the same mistake twice!

If most of your driving is around a town or city, you will find the performance of your electric car is great. Acceleration from a standing start is usually quick and the cars can be fun to drive in and around town.

If you have bought a *Neighborhood Electric Vehicle*[10] (NEV) in the United States, then your car can only be used in and around towns and cities. Your top speed on an NEV is limited but, even so, you will find that most of the time you are keeping up with the rest of the traffic flow around you.

If you have bought a used electric car, you may find that some much older models of electric car are sluggish when going up hills and that their performance is limited when driving out of town. This is rarely a problem with the new generation of electric cars that have become available since 2010. These will drive along at a reasonable speed both in town and out on faster roads.

---

[10] See our chapter on *Electric Cars You Can Drive Today* on page 35 for more information on the different categories of electric cars.

Speed and range

As with any other car, economy figures depend on how you drive it. If you drive everywhere as fast as possible, you will not travel as far on a single 'tank of electricity' as you will if you drive economically.

Other factors also make a difference to range. Running heating or air conditioning will make a difference (although some electric cars have a diesel-powered heater) but, less obviously, ambient temperature can also make a difference. Batteries perform better in warm weather than they do in very cold conditions.

Using lights or radio in the car will make very little difference to range. These ancillaries use relatively small amounts of energy.

Some drivers find they have to adjust their driving styles and techniques in order to achieve a decent range. Constant rapid acceleration and heavy braking, in particular, have a big impact on range and are best avoided if you are planning on driving a fair distance. These adjustments are not difficult and many people adopt them without even being aware that they have done so.

During the first few weeks, most drivers experiment with different driving styles to see how much they can improve the range of their car. Even if they do not need to drive far, many drivers feel a sense of achievement by extending their range.

As with fuel economy on any type of car, the biggest single difference you can make to range is to adjust your speed. The faster you go, the greater the wind resistance and the shorter the distance you will be able to drive. Conversely, the slower you travel, the further you will be able to go.

The calculation is not linear but generally, if you reduce your speed by 10%, you will be able to increase your range by around 15%.

The other big difference with electric cars is braking. Adjusting your driving technique to improve your braking can produce a big improvement on range:

- In combustion engine cars, a huge amount of kinetic energy is lost when you apply the brake pedal. The energy is converted to heat through brake friction, and is lost.

- In an electric car, regenerative braking uses the speed of the vehicle to power the motor, which in turn generates electricity to recharge the batteries.

- In many electric cars, regenerative braking can handle most of the braking effort required in day-to-day driving. I recently heard of one electric car owner whose twenty-five-year-old electric car is still on its original brake pads!

- In a city environment, using regenerative braking effectively can increase the range by up to 30%.

To get the best out of regenerative braking on an electric car, you tend to have to brake a little earlier than you normally would and brake gently rather than hard. Naturally, it takes time for a new electric car driver to get used to regenerative braking and to learn how to use it as effectively as possible.

Freedom from the service station

After driving an electric car for a few weeks, you start to appreciate that every time you get up in the morning you have a car with a 'tank full' of electricity and the freedom to drive wherever you want to during that day.

Suddenly, the benefits of being able to recharge your car at home rather than having to drive to a service station to refuel become apparent. No longer are you worrying about range; rather you are seeing the benefits of always having a car with a tank full of electricity every morning and never having to pay for fuel at a service station.

I recently lent a Nissan LEAF to one of my friends for a week. A confirmed petrol-head who was unconvinced about electric cars, he told me that after using the LEAF for a few days he understood how great it was to have a 'full tank of electricity' at the start of each day and how inconvenient it was to drive out of his way to buy fuel from a service station in his own car. For him, the mental switch from worrying about range to understanding how liberating an electric car could be took just three days.

Borrowing a 'plug full' of electricity

At some point during your electric car ownership, you will ask a friend if you can charge up your car while you are visiting. Most people are more than happy to lend a power socket. Many people will offer you a charge up when you turn up with an electric car without being asked. Friends are often quite surprised if you do not need the charge and turn them down!

The cost for the electricity you're borrowing is likely to be in the region of 10–20 cents per hour in the US, or 15–25p per hour in the UK during peak times.

If you use your electric car for business, you will often find that businesses are more than happy to offer a charge-up while you are visiting them. It is always best to phone up and ask first, to make sure that it is convenient.

If you are visiting a remote pub, restaurant or even an independent hotel in the evening, you will find that most of them are happy to offer a charge-up in return for your custom. But once again, I recommend you phone up and ask first; do not just turn up and expect them to accommodate you.

Camping and caravanning sites have onsite electricity, although if you are in Europe, you will require a 16 amp industrial plug in order to connect to it. Many sites have been happy to offer electric car charging for a small fee when requested.

If you want to borrow electricity in this way, make sure you have a suitable extension lead with you that can take the current required for charging up an electric car. The lead should not be coiled during charging as this can create a dangerous heat build-up in the cable. Make sure, also, that the lead has an RCD protected plug that switches off if there is an earth fault or a short circuit and that the socket to plug your standard car-charging cable into is protected from the elements.

Charging at work

Many electric car owners make arrangements with their employers to allow them to charge up their cars at work. Many employers are happy to provide this freely, as it portrays the company as being environmentally friendly. In some cases, the electric car owner has to pay their employer for the electricity used.

Quite often, an external power socket will need to be installed in order to allow an electric car to be charged up regularly. The cost of this will vary from site to site, but so long as you can park close to an external wall, it is rarely expensive. Alternatively, companies can buy and install charging-points for their staff car parks. Many of them are now doing this, often to help demonstrate their commitment to a more environmentally sustainable future.

Electric car charging-points

In many countries, more and more towns and cities are now installing electric car charging-points. Charging-points are being built into retail shopping areas, car parks and roadside parking facilities.

Around the world, local government and town councils are under pressure to make electric car charging-points available. The pressure is coming from politicians, environmental groups and car manufacturers, and from many electricity companies too. Charging-post solutions are available from several post manufacturers, and many new inner-city developments are being specified with electric car charging facilities.

Businesses are offering charging facilities for customers, too. Restaurants, pubs, hotels, shops and service stations are starting to offer charging facilities, often free of charge for customers and at a small fee for other owners.

The United Kingdom is leading the way, with charging facilities in virtually every major town and city around the country and with thousands of new charging locations planned for the near future. Car dealers have installed high-speed charging-points, whilst there is a nationwide network of rapid charging-points at motorway service stations that allow electric car owners to charge up their cars in around 25–35 minutes. This makes charging up easy, convenient and very inexpensive. Turn up, plug in, get a cup of coffee, and by the time you've drunk it, your car is charged up and ready to go.

Long-distance driving

Driving very long distances is now possible with an electric car, thanks to the network of rapid charging-points now in place in many countries.

In January 2011, the BBC demonstrated that long electric car journeys were not possible, attempting to drive a MINI-e electric car from London to Edinburgh (a distance of around 415 miles) and plugging the car into low-energy power sockets to recharge along the way. They managed to complete their journey, but it took four days! Three years later, in January 2014, actor Robert Llewellyn carried out the same journey using public charging-points. Travelling on motorways at an average speed of 68 mph (109 km/h), the journey was completed in less than thirteen hours.

Thanks to the vision of several private companies and backed by support from the government, the United Kingdom has one of the best charging networks in the world, with a network of high-speed charging-points operating in cities, car dealerships and at motorway and A-road services across the country. This is backed up by lower-speed charging-points, most of which are installed at car parks or at on-street parking points in the town or at shopping malls and business parks. Thanks to current government incentives in the United Kingdom, many of these charging-points can be accessed free of charge, allowing you to travel very long distances in an electric car with absolutely no cost for recharging on your journey.

High-speed rapid charging-points are terrific for people who drive longer distances only a few times each month. You can drive to your destination, stopping to recharge at services on your journey as you need to. In practical terms, it means stopping approximately every hour to plug in, and each stop takes 25–35 minutes.

If you are travelling long distances almost every day, you are very quickly going to get frustrated with this delay. If you are doing it once a week or less, you may find the inconvenience of having to stop less of a problem.

Personally, I find that the enforced stop allows me time to stretch my legs, get a cup of coffee, check my e-mails and even get some work done. If I am travelling with my children, the stop allows them to get out of the car, use the loo, grab a coke or a pizza, and we play a quick game together as a family. They enjoy the break and it has eliminated the previously inevitable question of "Are we nearly there yet?" when I take them on the 140-mile journey to see their cousins or grandparents.

Long term ownership

In writing this book, I surveyed several hundred existing electric car owners from all around the world, finding out about their experiences.

There have been electric cars on the road now for a number of years. Many owners have now owned their vehicles for four years or more, and a number are on their second or third electric car. Like me, there are a few owners who have used electric cars for the past eight to ten years.

When asked what they like about their vehicles, many owners talk about the lack of stress when driving. The vehicles are easy to drive, very smooth and quiet. Owners report that this make driving a much more pleasurable and calming experience.

Many owners also talk about the cost savings of driving an electric car, as well as the convenience of being able to charge their cars at home and never having to go to a service station for fuel.

Range is hardly ever mentioned with long-term owners. It quite simply is not regarded as an issue. In some cases, the electric car is a second car and therefore is not used for long-distance journeys. In other cases, the owners do not travel long distances by car at all, using the train instead. Other owners hire a conventional car or belong to a car club for the rare times they need to travel further.

There are many cases where electric car owners do not have access to a good nationwide charging infrastructure. Interestingly, the lack of a nationwide charging infrastructure is not regarded as an issue in these areas, although it is also true that where the infrastructure does not exist, electric cars tend to be used as second cars. In these areas, many owners see very little need to have a charging-point network at all.

As long-term electric car owners have discovered, the range deteriorates as the batteries get older. For the first four or five years, there is very little noticeable difference, but beyond this the range does start to reduce by a small amount each year. Lithium electric car batteries are expected to last as long as the car, but the range typically drop by 20% after around ten years of use, and will continue to decrease as the batteries age further.

Vehicle manufacturers are now offering longer battery guarantees on their latest models and claim that the battery degradation issue is less of a problem than it used to be. In a few cases, the batteries are being guaranteed for up to ten years to safeguard electric car owners from unexpected battery maintenance costs.

In cold weather, almost all owners have reported that their range can also be adversely affected, dropping by up to 30% in near-freezing conditions.

When questioned, the vast majority of people who have owned an electric car for two years or longer are so pleased with it that they intend to buy another electric car when they replace their existing model.

Chapter summary

- Owning and using an electric car is a new experience. It takes times to adapt.
- There is a lot to like about an electric car: the smooth acceleration, the lack of engine noise or vibration, the lack of pollution from the car itself, and the fun factor.
- New owners usually suffer from range fixation. This is overcome with experience.
- Nearly everybody forgets to plug the car in once. Very few people forget twice!
- There are various options available for charging up the car when travelling around. Some people use this to travel surprisingly long distances in a day.
- Very few long-term electric car owners regard range as an issue.
- Most long-term owners plan to buy another electric car when they replace their existing car.

Will an Electric Car Work for Me?

There is no doubt that more people are considering replacing their current car with an electric model. Leading UK motor industry experts Glass run surveys every six months about car-buying habits. In a poll carried out in the UK during January 2011, 53% of car owners said they would now consider an electric or hybrid-electric car as their next vehicle. Six months earlier, when the previous survey was run, only 9% of car owners said they would consider an electric or hybrid-electric car.

In the United States, the interest in electric cars is also huge. In a study conducted by IBM in the US during January 2014, nearly one quarter of all car buyers said they were either "likely" or "very likely" to consider purchasing an electric vehicle when the time came to replace their current car.

It is important to think about whether an electric car is the right choice for you. You need to consider whether it is a practical option and to understand your motivation for considering an electric car.

This process may take you some time to go through. Do not expect to have all the answers immediately. Owning an electric car is often a lifestyle choice, and deciding whether it is suitable for you can take some time.

Read through this chapter, consider it, and read the rest of the book. Then come back to this chapter and re-read it when you are ready. You may come up with an entirely different set of answers the second time around.

To understand whether an electric car will work for you, ask yourself some questions:

Regular long-distance driving

Every week I talk to people who love the idea of owning an electric car, but at the same time are regularly travelling long distances. This is often for work purposes, where their work can take them anywhere and everywhere with very little notice in advance.

The range of electric cars are improving all the time. When launched in 2010, the Nissan LEAF, for example, had an official range (based on EU figures) of 100 miles, or

160km. Today, the LEAF has a range of up to 150 miles (240km). The Chevrolet Bolt, being launched later this year, has an estimated range of over 200 miles (320km), whilst the Tesla Model S and Model X cars have a range of between 230 and 350 miles, depending on the model purchased.

There is, of course, a difference between the official range and the true range of a car. I can drive over 100 miles in my 2014 Nissan LEAF, for example, but only if I am driving carefully in city centre driving. If I am driving on faster roads between cities, my range drops to 70 miles.

So if you are driving long distance journeys on a daily basis, you need to either make sure that the car you choose is capable of doing these journeys, or you need to factor in recharging. I know many people who commute longer distances with a Nissan LEAF or similar, and this can be very practical and cost effective, especially if you can either find a public charging point or arrange for a charging point at your place of work, so you can recharge your car for the return journey.

Some countries, such as the United Kingdom, now have a nationwide high-speed rapid electric car charging infrastructure, where you can charge up your car in around thirty minutes. High-speed rapid charging-points are great for people who occasionally need to travel longer distances and therefore will not mind having to stop on their journey for a recharge. Many owners, including me, use these rapid charging-points once or twice each week and find them a practical solution for longer journeys. However, the downside of this is that journey times are inevitably longer than they would otherwise be. Whilst that can work for some people, others may find it tiresome.

What benefits will I get from owning an electric car?

There are many benefits in owning an electric car. Yet it is important to identify why *you* want to own an electric car. Ask yourself, what is important to you?

- Lower running costs?

- Better fuel economy?

- Reducing your environmental impact?

- The opportunity to drive something different, new and interesting?

You probably already have some definite ideas about the benefits of electric cars, but you may not yet know all the benefits of owning one. Reading this book should fill in some of the gaps and prompt some fresh ideas.

Where do I live and where do I drive?

If you live in a town or city and most of your driving is in built-up areas, an electric car makes a lot of sense. Almost every model of electric car on the market today will be suitable for this sort of travelling.

If you live in a small country village miles from anywhere, an electric car still makes sense, particularly if there is no service station nearby for putting fuel into your current car. You'll probably be more reliant on your car and travel longer distances than a 'townie', which means that you are more likely to be caught out by unplanned journeys. For this reason, you should choose an electric car that can be recharged relatively quickly at home: a 3–4 hour recharge time means that you are less likely to be caught out with a flat battery than an electric car with an 8 hour recharge time.

If you do a lot of travelling on high-speed roads for long distances, you will need to check that the electric car you choose is suitable, both in terms of performance and range. Range is considerably reduced at higher speeds. An electric car with an advertised range of 100 miles (160 km) is not going to achieve that range if travelling at 70–80 mph (112–128 km/h) for long periods of time.

How far do I travel in an average day?

Unless you have actually measured how many miles you travel on a daily and weekly basis, you are probably not aware of all the different journeys you do in your car and how far you drive.

People who claim to know how many miles they travel each day are often surprised when they actually measure their journeys, as opposed to estimating them. Most people over-estimate their journey lengths by between a quarter and a third and forget to take into account one fifth of all their car journeys.

Keep a diary for a couple of weeks, noting when you use your car and how far you travel each time you use it. Also, make a note of the type of roads you are driving on and your average and top speeds. The results could surprise you and will give you accurate figures of what you really need your car to do for you.

Once you know how far you travel on a daily basis, you should choose an electric car with a range that is at least double the range than you expect to need.

For instance, if you need to travel 25 miles (40 km) a day and do not have the facility to recharge the car during the day, you need to choose a car with a range of at least 50 miles (80 km).

Having this extra range means that you should comfortably be able to travel as far as you need each day, no matter what happens. It means you won't worry about using the heating or the air conditioning, nor are you likely to be caught out by unexpected journeys.

Range – the true story

Real world range is difficult to pin down. It depends on driving style, road conditions, weather and temperature, how well the car is maintained and how much weight the car is carrying.

Unfortunately, the official 'New European Driving Cycle' (NEDC) tests that are carried out in Europe are so unlike the real world that they provide an entirely unrealistic guide for consumers wishing to know the true range of an electric car.

The United States, however, has adopted a far better model for range, with a series of tests designed by the Environmental Protection Agency (EPA) to give consumers a much clearer idea of the efficiency of their car and the range that they could expect to achieve. Their tests use a combination of city driving and highway driving cycles, whilst using a certain amount of air conditioning or heating where required. These figures mean that the ranges quoted in the US are far more realistic than those quoted in Europe. Indeed, in many cases it is easy for owners to beat the US figures, depending on how they use their cars.

Incidentally, it is worth noting that a new worldwide set of test procedures is currently being developed which will replace both the NEDC and EPA tests. Called the 'Worldwide Harmonized Light Vehicles Test Procedures' (WLTP), these tests are designed to reflect real-world driving far more accurately than either the NEDC or EPA tests, and will start being used in 2016.

However, there is only one way to ascertain real-world range, and that is to try it yourself. As part of the research for this book, I carried out a number of test journeys in order to ascertain the realistic range in two electric cars: a Nissan LEAF and a Mitsubishi i-MiEV.

The Nissan LEAF is one of the best-selling electric cars in the world. Since its launch in 2011, over 200,000 have been sold and the car is now produced in Japan, the United Kingdom and the United States. It is a family-sized car with seating for five and has an official range of 124 miles (200 km) under the NEDC tests and 75 miles (121 km) under the EPA tests.

The Mitsubishi i-MiEV is a far more compact car than the LEAF. It is a small sub-compact city car, capable of taking four people in comfort. According to Mitsubishi, the car has an official range of 93 miles (150 km) under NEDC tests and 62 miles (100 km) under EPA tests. The car is also sold as a Peugeot iOn and a Citroen C-Zero, and has become a popular choice in Japan and across Europe.

A series of tests was carried out at different times of the year using two different drivers to ascertain the 'real-world' range of the electric car. The cars were both driven on public roads in an ambient temperature of 17–20°C (62–68°F), without the use of heating or air conditioning.

It is worth noting that there is now a longer range Nissan LEAF available with a significantly greater range (150 miles under European NEDC testing). Whilst the range figures on the new model are higher, the real world range would drop in a similar way to the older car.

	Nissan LEAF	Mitsubishi iMiEV
Official range (European NEDC tests)	124 miles (200 km)	93 miles (150 km)
Official range (US EPA tests)	75 miles (121 km)	62 miles (100 km)
City centre driving	98 miles (156 km)	91 miles (147 km)
Urban/extra urban	82 miles (131 km)	79 miles (126 km)
Cross country roads	74 miles (118 km)	75 miles (120 km)
Cross country roads – fast driving	48 miles (77 km)	46 miles (74 km)
Dual carriageway/freeway eco driving	69 miles (110 km)	68 miles (109 km)
Dual carriageway/freeway normal driving	52 miles (83 km)	51 miles (82 km)

In the winter, the range of the Mitsubishi i-MiEV decreased by around 30%, dropping to nearer 45% for heating and defrosting the car. In the Nissan LEAF, range decreased by around 10%, dropping to nearer 30% for heating and defrosting the car.

As you can see, real-world figures differ dramatically from the official figures, particularly when driving at higher speeds and particularly when compared to the European NEDC tests. Around town and in urban areas, the range of the electric car remains good. When driving at higher speeds, particularly if driving energetically, range decreases dramatically. If you are expecting to achieve the official European range figures in regular driving on a daily basis in real-world driving conditions, you are very likely to be disappointed.

Where can I plug in?

Do you have a garage, or at least off-road parking that allows you to plug in your car to charge it at night?

If you do, charging up your electric car is going to be easy. You may wish to install an outside power socket or a dedicated electric car charging unit to speed up the car charging but, in essence, you are going to have no difficulties charging up your car.

If you park in a private communal area, then you may need to seek permission from whoever manages this parking area and you will almost certainly have to pay for any work to be carried out.

Often you will get an extremely co-operative response when it comes to arranging this. If you can fit a charging-point to an outside wall and there is a suitable power supply at hand, then the costs can be reasonable. If, however, you need to install a freestanding charging-post and run underground cables, the cost can quickly become very significant.

If you only have on-road parking, you may not be able to charge your electric car at your house. Some electric car owners have been known to trail cables across footpaths, but this is very dangerous as it poses a significant trip hazard for young children and the elderly.

I have seen one ingenious method for roadside charging, where an electric car owner ran a cable from their house into a tree at the edge of the road. They then fitted a charging socket inside the tree and cleverly disguised it all to look like a bird's nest!

In some countries, local council offices can arrange for an electric charging-point to be installed outside your house, installing a *power bollard* by the side of the road. Costs vary dramatically and are rarely low, although subsidies for installing household power bollards are being considered in many countries.

Some houses with no off-road parking do have a small yard at the front of the house. These are often not large enough for a full-sized car, but many compact electric cars are small enough to be parked on these yards. You can arrange for your local council to install a 'drop kerb' on the footpath next to your house and convert the front yard into a short driveway, thereby ensuring that you always have a charging bay for your electric car.

Some electric car owners have made arrangements with local businesses, allowing them to charge up at the business premises outside business hours. There are benefits to the business in allowing this:

- There is activity at their premises outside working hours, thereby making the property less of a target for burglary and vandalism.

- It can help local businesses to nurture goodwill and a reputation for being environmentally friendly.

Costs for installing a dedicated charging-point at home

Assuming that you have a suitable electrical connection nearby (which means 240 volts in most cases), there are a number of different options for installing a dedicated charging-point.

While it is not the fastest charge option, an electric car can be charged up from a standard domestic power socket. There are some caveats, however. Charging an electric car requires a constant high current, which can heat up the plug and even weld the plug to the socket if it remains plugged in constantly. Make sure you disconnect the cable from your mains socket after each charge. If you use an extension lead, use it with care: ensure that the lead has a high enough amp rating and that it is entirely uncoiled to avoid overheating. You can purchase an external power socket that can be fitted to an outside wall from your local builder's merchant. Look for a socket with IP54 environmental protection to ensure that it can be used in all weathers.

If you are based in North America and only have access to an 110v power supply for charging, charging from a domestic power socket is not usually a practical solution. Most electric cars will take 15–20 hours to recharge from an 110v power socket, which is far too slow to make it a practical day-to-day choice.

A far better solution than a standard power socket is to have a dedicated electric vehicle home charging-point installed. These provide a higher current, allowing you to charge up an electric car faster. How fast depends on the car. In the case of the Ford Focus Electric, a BMW i3 or a Nissan LEAF, a dedicated home charging-point can charge up the car in around 3–4 hours, while you can charge a Mitsubishi i-MiEV or a Chevrolet Spark EV from a dedicated home charging-point in around 6 hours.

Costs for these home charging-points vary. In the United States, a charging-point purchased from *Best Buy* will cost around $1,000, including installation, while charging-points from other companies will cost between $300 and $1,500. Most car dealers who sell electric cars also offer electric car charging-points and you can usually negotiate a discount on your charging-point if you order it at the same time as purchasing your car.

Above: A wall-mounted charging point that can be installed on an external wall at your house by a qualified electrician.

In the United Kingdom, there is currently a grant scheme in place, covering up to 75% of the cost of installing an electric car charging point at your home. The scheme allows you to apply for a charging-point and have it fitted by an approved installer.

Various companies are offering this service, but it is worth shopping around to find the best provider. British Gas, for example, offer the service, but the charging-points are actually installed by a company called Chargemaster. Some EV owners who have experienced problems with their charging-points find that British Gas struggle to support them. If you go direct to Chargemaster yourself, you usually get a faster service and better after-sales support. Pod Point is another well-known UK-based charge-point company who offer a very smart wall-mounted charging-point, whilst Rolec also offer a home charging point with a choice of posts and bollards, allowing the charging point to be installed by a driveway rather than mounted to a wall.

Shop around

If you are considering a dedicated charging-point, it is worth shopping around for the best deal. Many electricity suppliers now offer electric car charging-points, as do car dealers, but if you are paying for your charging-point, look around at what options are available to you. You are not forced to buy the charging-station sold by the manufacturer of your car.

Most home electric car charging-points are available in either 16 amps or 32 amps. Most of the latest electric cars have the capability to charge up at 32 amps, so if you have the choice and the cost differential is not too great, go for the higher charge rate option.

Charging cable standards

Home charging-stations either come with a cable that allows you to plug it straight into your car, just like a fuel pump, or with a socket which you use with the cable from your car.

The charging-stations with the built-in cable are great for convenience, but do have a drawback: they are not universal. There are two different standards for electric car charge cable connectors and unfortunately the socket on cars such as the Nissan LEAF and Mitsubishi i-MiEV are not the same as the socket on the BMW i3 or VW e-Golf.

This means that if you buy a charging-station with a built in cable, you may need to upgrade it when you buy your next electric car. This is not a problem if you choose a charging-point that incorporates a power socket, as the power socket connector is universal.

Can I charge my car elsewhere?

If your place of work has a private car park, you may wish to enquire whether you would be allowed to charge your car at work, either by trailing a power lead through a window on an occasional basis, or having an external power socket installed for more regular use.

Even if you live well within range of your workplace, having the ability to charge your car at work can be useful from time to time.

Many towns and cities now have public charging-points, with new ones becoming available all the time. Knowing the location of your nearby charging-points is useful information, even if you do not plan to use them. If nothing else, knowing where they are gives you peace of mind.

You will need to make sure that you have a suitable charging cable in order to charge at most public charging-points. Usually these are supplied when you buy the car. If not, make sure that you order one from the options list when you place the order for your car!

Can I remove the batteries to recharge them?

Many people assume that an electric car battery would be similar in size and weight to a standard starter battery used in a conventional car.

Sadly, this is not the case. The batteries required to power an electric car are bulky and heavy. Imagine a battery pack at least the size and weight of a large combustion engine. You would require heavy lifting gear to remove the whole pack in one go.

How often do I need to be able to drive further than an electric car will allow me to go?

From time to time, most people will need to travel further than an electric car will allow them to go. If you are buying an electric car as a second car, this is not a major consideration. Instead, you can simply use your other car for the long-distance journeys.

If you are buying an electric car as an only vehicle, there are options available to you:

- Stop en route to charge up at a rapid charging-station.

- Travel by train.

- Join a car club that allows you access to a car as and when you need one and allows you to hire by the hour.

- Hire a car from a car hire company.

- Some electric car suppliers have arrangements with car hire companies to allow you to hire cars for occasional use at discounted rates.

Stopping en route to charge up at a rapid charging-station can work well. If you are based in the United Kingdom and drive a Nissan LEAF, for example, there are rapid charging-stations all around the country, making it possible to drive virtually anywhere. The same can be said about California. Other locations are not so well covered and there are often big gaps in the infrastructure.

Hiring a car for occasional use actually makes a lot of sense. It allows you to choose a suitable car for the purpose: a large car for travelling with a group of people, a small car on another occasion; even a luxury car for impressing your boss.

Sharing a car with a car club can cost as little as £4.50 per hour in the UK, or $8 per hour in the United States. The cost typically includes all fuel costs and insurance. A full day's car rental can cost as little as £25 in the UK, or $30 in the United States.

If you are regularly going to be travelling further than an electric car will allow you to go and you do not have access to another car, are you going to be happy to use public transport? Is this going to be practical? Regularly commuting from one city to another by train may be a practical option for some people but may be completely impossible for others.

What if an electric car is not suitable for me now?

This is a good point to take stock. If you have already decided against an electric car, then that is a shame, but at least you have vital information. It is better than spending thousands on a car and then discovering that it is not suitable for you.

Electric cars are continuing to evolve and improve. Even if an electric car is not a practical solution for you now, it may well be in two or three years' time.

There two other options that you may wish to consider:

Sharing a car

Do you live close to friends or family who would also be interested in owning an electric car? If so, why not pool resources and buy an electric car between you? If you then share all the cars you have, you can use an electric car for shorter journeys and another car for long-distance driving.

Communities can also club together to buy and run an electric car. The car is parked in a permanent location such as a local community area or library, where the car can be kept on charge. The keys are held in a wall-mounted safe with combination lock, and booking is set up using one of the many freely available internet-based diary systems. When somebody in the scheme wants to use the electric car, they simply book it, pick up the keys and drive away.

There are other ways of sharing a car, too. In the UK, Easy Car Club (www.EasyCarClub.com) allows you to register your car for car sharing, so you can earn money from your car while it is not in use.

Sharing a car may sound radical, but it is gaining momentum, with schemes in many major cities. In big cities such as New York, Chicago and London, car sharing is now commonplace, with companies such as Zipcar, Car2Go and Hertz all offering car-sharing schemes. In the UK, over 140,000 people have now subscribed to car-sharing clubs.

Change your life!

This might sound rather drastic, but many people who are looking to buy an electric car are doing so as part of a much larger, life-changing transition. It is not uncommon for people to buy electric cars, having recently changed jobs, retired or moved house, as part of a bigger plan to improve their lives.

Your dream life may consist of living in a cave in the side of a hill, living in the latest ultra-high tech eco-home, or just living a simpler life. Whatever it is, learn how you can reduce your dependence on your existing car first. Do not try to live your life with an electric car before you are ready.

Chapter summary

- For some people, an electric car is the perfect choice. For others, it is impractical, at least for the time being.

- You need to consider how you use a car.

- Owning an electric car is often a lifestyle choice. Do not hurry the process.

- There are a few practical points you need to consider:

    o Where you live and how far you travel.

    o Where you can plug in to recharge.

    o How often you need to travel beyond the range of an electric car and how you plan to do that.

Purchasing and Running Costs

In general, electric cars are more expensive to purchase than comparable combustion engine cars. This price differential is dropping, but as a rough rule of thumb, electric cars tend to cost about 15–20% more than conventional cars.

These costs are offset by the significantly lower running costs of an electric car. Specifically, the cost of fuel is significantly lower, quite often entirely offsetting the additional cost of purchasing the car in the first place, particularly if you are leasing your car.

Leasing schemes and purchasing plans

Some manufacturers only provide their electric cars on leasing schemes, as opposed to outright purchase. Citroen and Peugeot, for example, are only offering their cars on fixed-term leases, after which the cars are returned to the manufacturers.

The leasing cost for these cars is often higher than the leasing cost for an equivalent combustion engine car. However, almost all of these leasing schemes incorporate servicing, warranty and vehicle breakdown, and in some cases insurance. Remember, neither are you paying for fuel from a service station. All this helps offset the cost difference.

In Europe, the price of leasing an electric car is falling rapidly. At the beginning of 2011, the Nissan LEAF cost around £400 ($615) per month to lease over five years, plus an initial deposit of £2,400 ($3,800). Today, you can lease a Nissan LEAF for £100–200 ($160–$320) per month with an initial deposit of £2,000 ($3,200). The same is true for other electric cars. Renault, BMW and Mitsubishi all offer electric cars at competitive lease prices. In most cases, an electric car will only cost a little more than the equivalent combustion engine car to lease.

In Europe, Peugeot are offering an interesting twist on conventional car leasing, with the release of their new Peugeot iOn electric car. Customers lease the car over a four-year period for a fixed monthly cost incorporating servicing, maintenance and use. As part of the scheme, owners can temporarily swap their electric car for a conventional car should they need to travel longer distances or simply require a larger car.

For some people, the point has been reached where it is possible to lease and run an electric car for less than the monthly cost of pumping fuel into a conventional car. I have reached this point with my current car (*see below*). As fuel prices continue to rise and more owners work out that they can save money with an electric car, it is predicted that a lot of people will ditch their gas-guzzlers and spend their fuel money on leasing and running a brand new electric car instead.

Reaching the Tipping Point

I lease my current car, a 2015 model Nissan LEAF. After paying a £1,000 deposit, I am paying £220 a month for four years, driving 25,000 miles a year. The amount of electricity I use at home to recharge the car costs me £24 each month.

At the time of writing, one litre of petrol costs between 99p and £1.10. If I drove a conventional car and got 40 miles to the gallon, the cost of a month's fuel at today's prices would be in the region of £248.

Because I am not paying for fuel, I am paying £4 less each month than a petrol bill to both lease and fuel a brand new car.

Government subsidies and discounts

Various countries have subsidy schemes in place to reduce the purchase price of an electric vehicle. In the United Kingdom, for example, all electric cars qualify for a 35% discount, up to a maximum of either £2,500 or £4,500 (depending on the model), whilst many electric commercial vehicles qualify for a 20% discount, up to a maximum of £8,000.

In the United States, electric car owners are eligible to claim a tax credit against the purchase cost of their electric vehicle. The amount of tax credit varies depending on the size of the battery in the car, but is a minimum of $2,500 and a maximum of $7,500. You can also claim a tax credit of up to $1,000 on installing your own electric car charging point at your home. Unlike the British scheme, the tax credit works as a tax rebate, which means customers have to pay the full price for the car and claim the discount later.

Many US customers also benefit from state-level tax incentives. California, Texas and Rhode Island, for example, offers a further $2,500 tax rebate for new electric car purchases, whilst in Colorado, up to $6,000 is available in tax credits against the purchase of an electric car.

In Canada, residents in British Columbia and Ontario can also receive tax credits for $5,000, against the purchase of electric cars.

Buying second hand

Based on the US Kelly 'Blue Book' and UK's Glass's Guide, used Nissan LEAFs and Mitsubishi i-MiEVs retain 33–38% of their new value at three years old, based on an average mileage and condition. This makes them an excellent buy as used cars.

Exact values do vary, depending on where the cars are based. In London, for example, where electric cars are exempt from the daily congestion charge, used prices are around 8% higher than the rest of the country. The same is true in California, where used electric cars are sold at higher than average prices, thanks to the state-wide charging network and higher public acceptance of electric vehicles.

A good selection of used cars can typically be found at franchised dealers. You will also find them advertised online by independent dealers and private owners.

Fuel costs

The biggest single saving with an electric car is a reduction in fuel costs. Instead of paying for fuel at a service station, an electric car owner simply plugs a cable into the car overnight and charges up using off-peak electricity.

From empty to full, most electric cars use between 8 and 24 kWh of electricity (1 kWh = 1 unit of electricity) to provide a complete charge. In the UK, off-peak electricity costs around 8 pence per kWh. In the US, off-peak electricity costs around 11 cents.

Here is an example. If an electric car driver travels around 30 miles (48 km) a day, that equates to around 900 miles (1440 km) a month.

If your car is parked overnight, the car can be charged using off-peak electricity. Thirty miles of driving a typical electric car would equate to around 6 kWh of power.

Cost savings in the United Kingdom

In the United Kingdom, based on 8 pence a unit cost, the cost of recharging the car would be 48 pence a night, or £14.40 a month.

Compared to a conventional car that gives 35 mpg (7.7 km per litre, based on the imperial gallon), you are using roughly 136 litres of petrol per month to go the same distance. At a cost of £1.30 a litre, that is a total cost of £177 per month.

In this example, the cost of charging an electric car is around 8% of the cost of putting fuel into a conventional car.

Real life cost savings in the United Kingdom

I have been using an electric car for many years. At the end of 2009, I installed a separate electricity meter in order to measure how much electricity I used, and to measure the true cost of that electricity.

During 2010, my main family car was an electric Mitsubishi i MiEV. Over a period of a year, I drove 8,600 miles. The total cost of that electricity was £85.60 ($131). That calculates to a monthly cost of £7.13 ($10.10).

In March 2011, my main family car was an electric TATA Indica Vista EV, which I used to drive 11,400 miles. The average cost of the electricity per month was £10.90 ($15.40).

Today I drive a Nissan LEAF and it is the only car in our family. Due to a change in job, my annual mileage is now around 25,000 a year. My monthly electricity bill for charging my car is around £24 ($34).

Cost savings in the United States

In the United States, based on 11 cents a unit cost, the cost of charging the same car would be 66 cents a night; just under $20 a month.

Compared to a conventional car that gives 30 mpg, you are using roughly 35 gallons of petrol per month to go the same distance. At a cost of $3.15 per US gallon, that is a total cost of $110 a month.

In this example, the cost of charging an electric car is around 18% of the cost of putting fuel into a conventional car.

Servicing costs

One benefit that has been claimed for electric cars is lower servicing costs. After all, fewer parts equates to simpler maintenance, so, in theory, servicing an electric car should cost less than a conventional car.

This is true up to a point. If you want to service your electric car yourself, the chances are, it will be cheaper than servicing any other vehicle. Many people who have older electric cars do the servicing themselves and have saved themselves money as a result. This is especially true if you have an older model electric car, such as a G-Wiz or a GEM, which uses lead acid batteries. Lead acid batteries need replacing every few years and this can be expensive.

If you want to have your car serviced by an electric vehicle dealer, you pay more for the specialist skills. Inevitably, this means that the overall servicing costs for an electric car are about the same as for other cars.

Chapter summary

- In general, electric cars are more expensive to purchase than other cars.

- Charging an electric car is very cheap, but if you are using one of the older electric cars that uses lead acid batteries, you should factor in a percentage of the cost for replacing the batteries every few years.

- Servicing costs are comparable with other cars.

- Resale values for electric cars are on a par with other cars.

Electric Car Charging Networks

There have been various surveys carried out around the world to look at what would be required before people were prepared to buy an electric car. In almost every study, the biggest single reason for not buying an electric car is the lack of a national network of electric car charging-points.

The first electric car charging-points appeared in California and London around fifteen years ago, when the first experimental electric cars, such as the GM EV1 and the Honda EV Plus in California and the Ford TH!NK in Europe appeared in small numbers. The charging-points allowed electric car owners to charge up their cars while parked, over a period of several hours. Most of these were neglected, and some were removed completely, when the vehicles failed to gain any real commercial traction.

London was one of the first cities in the world to get back its charging network, thanks to the success of the G-Wiz electric car in the mid-2000s. Charging-stations started appearing in car parks and this was extended by a number of small businesses and private individuals who helped build up a new network of charging-points to match the new demand from G-Wiz owners.

It was in 2009 and 2010 that the current public charging-points started to appear elsewhere. Like the earlier charge-points, these were low-speed charging-stations that allowed electric car owners to charge up their cars while they were parked. Charging times were measured in hours rather than minutes.

High-speed charging-points started to appear in 2011, when Nissan launched the LEAF. Early charging-stations were installed at Nissan dealers, but soon started to appear elsewhere as well. Charging from a high-speed charging-point originally meant the cars would only charge to 80% of capacity, but the charge time was reduced to between 25 and 30 minutes for most cars.

Today, it is true that many countries have a national charging network for electric cars. Unfortunately, not all charging networks are created equally... and not all cars can use all charging-points.

The United Kingdom has rapid charging points at every motorway service station in the country, thanks to the Ecotricity Electric Highway. Electric car owners can charge up for free in as little as half an hour.

Charging-point standards

Public charging-points come in different capacities:

- In the United States, low power or slow charging uses standard US domestic power sockets (referred to as Level 1 charging), which can take 15–20 hours for a full charge.
- Medium power, or standard charging (referred to as Level 2 charging in the United States and confusingly sometimes referred to as 'fast charging' in the UK), typically provides somewhere between 3 and 7 kW charging facility. A typical electric car will take around 6–8 hours to fully charge from a 3 kW charging facility and 3–4 hours from a 7 kW charger.
- High power, or rapid charging (referred to as Level 3 charging in the United States), typically provides either a 40–50 kW high-voltage charging facility, which can take 25–30 minutes to provide a full charge, or a 22 kW high-voltage facility, which can take around one hour for a full charge.

Low- and medium-power charging-points are typically found on streets in towns and cities or in public car parks, in places you would expect to park for a few hours before returning to your car. There are typically spaces for two or four cars to park and charge at each location.

High-power charging-points are typically found at motorway (highway) services, on major trunk roads or at car dealers. There is typically space for only one car to charge up at any one time, although charging sites are now expanding to cater for two to four cars at a time, as demand for high-power charging-points increases.

Charging-points at car dealers are often only available for owners who have bought the brand of car that dealer is selling: you can typically only charge a Nissan car at a Nissan dealer, for example.

High-power rapid charging-points come in two different types. 'AC' charging-points are suitable for cars that can accept a high-power AC charge, such as the Renault Zoe, and 'DC' charging-points are suitable for cars that can accept a high-power DC charge, such as the Nissan LEAF, the Chevrolet Spark EV and the Mitsubishi i-MiEV.

Unfortunately, it gets even more complicated with DC charging-points, as there are two different standards for DC charging, neither of which is compatible with the other:

- CHAdeMO is used by Nissan and Mitsubishi and is the most established, with the most rapid charging-points available
- CCS (also referred to as SAE Combo or Combi Charging), is just coming online now. It is used by the big American and European manufacturers, including GM, Ford, BMW, Mercedes, VW and BMW.

Left: the CHAdeMO connector.   Right: the CSS / Combi Charging connector.

Finally, if you have a Tesla, their high-speed rapid charging uses their own proprietary standard of supercharger, allowing them to add around 170 miles of range to their cars in just 30 minutes. However, Tesla owners also charge up at high-speed AC charging-points on motorways, or charge using a CHAdeMO socket with an appropriate adaptor.

This sounds confusing and it certainly does not help the prospective owner's confidence levels. The good news is that already, most high-power charging-points can support both AC and DC high-power charging, and charging network operators are working hard to ensure that all electric cars will work with all the charging-points they have available.

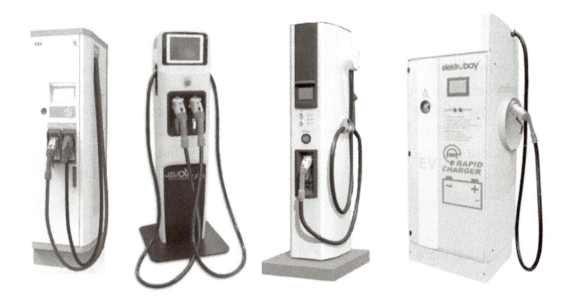

Above: Multi-charging stations that can provide high-power rapid charging to virtually all electric cars already exist and are being rolled out by network operators. In the United Kingdom, Ecotricity are upgrading their motorway network of charging stations to cater for the fast charging needs of all electric vehicle owners.

Why so many options?

There are genuine technical reasons for there being all these different standards, but that hasn't stopped the conspiracy theorists from speculating that 'big oil' is trying to make things difficult for electric car owners.

The reason for the different standards is that electric vehicle technology is improving rapidly. The demand for faster and faster charging means that CHAdeMO-based charging will reach its limits sooner than CCS. That is not an issue for electric car owners today, or even for electric car owners for the next five to eight years, but it does mean that at some point in the future, CCS is likely to become the main standard in Europe and America.

That said, owners of CHAdeMO cars are highly unlikely to be left in the cold without high-speed charging. Once charging-stations have been upgraded to work with CCS, they will be compatible with both CHAdeMO and CCS. Furthermore, most Japanese and Korean manufacturers have committed to the CHAdeMO standard, so the technology will be with us for many years to come.

Whilst it is frustrating that there are so many different standards, in reality it isn't much different to putting fuel into a conventional car. Go to any service station in Europe and you'll have a choice of unleaded petrol, super unleaded petrol, diesel and 'ultimate' diesel. Many service stations also offer LPG or CNG, some also offer biodiesel and there are even hydrogen fuel pumps appearing in some parts of the country. Yet somehow we all manage to fill our cars up from the correct pump without too many mishaps!

Teething Troubles

It is fair to say that there have been a lot of teething troubles with high-power charging-points. Early charging-points were notoriously unreliable, often failing with grounding faults in wet weather. These problems have largely been resolved. High-power charging-points are significantly more reliable than they previously were.

In the United Kingdom, I now use a Nissan LEAF electric car for all my driving, including a regularly driven 150 mile round trip. From personal experience, I know that the charging-point network is now reliable enough to work, with new charging-station sites appearing monthly and existing charging-stations being expanded to handle the demand.

If you are using a car with a CCS rapid-charging connector, the teething troubles are not yet over. This new standard is only just becoming available on cars, and the rapid charging-stations themselves need to be upgraded to work with them. This work takes time, and whilst they are starting to appear (in the United Kingdom, for example, they are currently being installed at the rate of four a day and there is now a nationwide CCS network in place, but it is not as comprehensive as the existing CHAdeMO network), other countries are still lagging a little behind.

That is not to say that rapid charging is entirely fault free. There are still a few occasions when a charging-point is not working, so if I am planning to use one on my route, I check online first to make sure the charge-point I want to use is in action.

The good news is that, inevitably, if one rapid charge-point is out of action, there is another one nearby. I regularly drive the 120-mile trip to visit my parents. On my usual route, I pass five rapid charging-points. If I am prepared to divert 3–5 miles from my usual route, I can get to a further seven. I have never been left stranded on a long-distance journey and been unable to charge.

Congestion

Unlike a fuel pump, where you can refuel in three or four minutes and there are always lots of fuel pumps at a single service station, rapid electric car charging-points can usually only charge up one car at a time.

Most sites either one or two rapid charging-points, and there have been occasions when I have turned up at a charge-point only to find it is already in use. Usually, this is not a problem. I check the charge-point to see how much charge time is left. If it is just three or four minutes, I'll wait and then plug in as soon as the other car is finished. If it is going to take longer I'll usually continue my journey and charge up at the next charging-point on my route.

The good news is that network operators are constantly adding new charging-stations, both in new locations and at existing sites. In the United Kingdom, at least, supply is comfortably matching demand.

What cars work with what charging-points?

Every electric car will work with the low and medium-power charging-points, so long as you have the appropriate cable. For high-power rapid charging, these are the options:

Type	Availability	Description	Vehicles
CHAdeMO	High availability	The original standard, backed by the Japanese car manufacturers. Heavy investment by Nissan and early private charge network investors means that this is almost universally available, with rapid charging-points.	Citroen C-Zero KIA Soul EV Mahindra e2o Mitsubishi i-MiEV Mitsubishi Outlander Nissan LEAF Nissan e-NV200 Peugeot iOn Tesla Model S and X *with optional adaptor*
CCS also known as SAE Combo or Combi Charging	US: Low availability UK: Medium availability	The standard adopted by most European and American car makers. Ultimately it is likely to become the most used, but is currently not available at all rapid charging-points. This will change. By mid-2016, truly universal rapid charging-points with CCS as well as CHAdeMO will be the norm.	BMW i3 Ford Focus Electric VW e-Golf VW e-Up Honda Fit EV

What if I want to buy an electric car today and guarantee fast charging?

If you want to buy an electric car today and be able to fast charge it at as many locations as possible, your best choices are to buy a car that can use a CHAdeMO connector or a rapid charge AC.

This will change. By mid-2016 in the United Kingdom and mid-2017 in the United States, it is likely that you'll be able to buy any electric car that is equipped with a rapid charge capability and use it at virtually every rapid charging-station.

How to use a public charging-point

In order to use a public charging-point, you need to subscribe to a charging network, which usually involves an enrolment fee. You will then be given an access card to identify yourself when you arrive at a charging-point.

There is more than one charging network in most countries. In the early days, this meant you needed multiple access cards, depending on which charging-point you were using, but thankfully this is less of an issue now. Most charging networks have cross-network agreements, so if you turn up at one charging-point you can still use it by means of an access card from a different charging network.

Once you have arrived at the charging-point, you identify yourself to the charging-point by holding the access card up to the card reader. Authentication typically takes 5–10 seconds. Then follow the instructions on the screen of the charge-point itself and plug the cable into your car.

If you are connecting into a low or medium-power charging-point, you will need to bring your own power cable with you. Some car manufacturers supply the appropriate cable when you buy your car, but with others the charging cable is an optional extra. Plug the cable into the car and then into the charge-point. The charging-point will lock the cable into the unit so that it cannot be removed, and will display on the screen the charging status.

If you are connecting into a high-power rapid charger, the cable is attached to the side of the charger (the unit looks very much like a fuel pump from a regular service station). Plug the cable into your car. The charging-point will then display the charging status.

Completing the charge is very similar. You identify yourself to the charge-point using your access card. This unlocks the cable so you can disconnect and drive away. Payment is typically carried out centrally, using the credit card details that you supplied when you set up your charging network account.

Charging-point networks in the United States

A number of private charging networks exist in the United States, but at present these tend to only cover the major cities and some interstates. A few states still have no public charging-points available whatsoever.

The main network providers are Charge-point (www.ChargePoint.com) and evGo from NRG (www.nrgevgo.com). Both companies provide a combination of high-power rapid charging and standard speed charging, with high-power rapid charging-points typically situated on long-distance routes and standard speed charging-points typically situated in shopping malls and in downtown locations.

Customers can either pay a monthly charge in order to use the charging-points or adopt a pay-as-you-go approach, which costs more per charge but is better for occasional ad hoc use. Nissan LEAF owners can take advantage of Nissan's 'No Charge to Charge' programme for free charging at many locations.

Charging-point networks in the United Kingdom

British electric car owners probably have the best charging-point options available today, many of which can be accessed entirely free of charge. The UK now has a complete network of high-speed rapid charging-points covering the entire country. Furthermore, thanks to extensive government funding, virtually all major towns and cities now have medium-power standard speed charging facilities, whilst Nissan have ensured that high-speed rapid charging-points have been installed in many of their dealers.

Most excitingly, the British energy company Ecotricity has installed a national high-speed rapid charging network and has charging-points at every motorway service station across the country, at many service stations on major trunk roads and within a growing number of city locations. The Ecotricity *Electric Highway* charging-stations can be accessed free of charge, and the company says they have no plans to introduce payment in the near future. EV drivers can register for an access card at *www.Ecotricity.co.uk/for-the-road* and simply use the charging-stations whenever they want.

The vast majority of standard-speed charging-stations in the country has been installed by one of two companies: Charge Master (www.ChargeMasterPlc.com) or POD Point (www.pod-point.com). Both companies operate a subscription scheme, where customers can either pay a monthly fee to access the network, or adopt a pay-as-you-go approach. The two companies 'share' charging-points, allowing you to charge up at any charging-point, whichever scheme you subscribe to.

Chapter summary

- Charging networks are in place and growing in number all the time.
- Standard-speed charging takes between 3 and 6 hours to charge your car, depending on the model you have.
- High-speed rapid charging takes around half an hour to charge your car, but there is more than one standard for rapid chargers.
- Rapid chargers using the CHAdeMO and AC charging standard are now commonplace, allowing cars that use this charging standard (predominantly the Nissan and Mitsubishi cars) to be freely used in many areas.
- The United Kingdom has a truly country-wide network of high-speed rapid charging-points, most of which can be used free of charge.

Electric Car Categories

The electric car industry is still young and full of potential. It has opened up many opportunities for fresh thinking and new ideas for transportation. In automotive design studios and in governmental think tanks alike, people are discussing what the car of the future will look like. Is an electric car simply a conventional car with an electric motor, or is it an entirely different type of vehicle? What about self-driving cars? How can car design influence society? How is society influencing car design? What opportunities does an electric vehicle offer that conventional cars cannot?

It's a fascinating subject, which I touch on in the chapter *Why are Electric Cars Important Today?*, starting on page 23. Companies are producing exciting concepts, many of which are a far cry from conventional cars. The Google self-driving car is currently on test in the United States, whilst the highly distinctive Renault Twizy electric quadricycle has been on sale in Europe for a number of years. Neither of these vehicles is legally classified as a car, but they offer intriguing insights into some of the ways that electric vehicles may change our thinking about road transport in the future.

Categories of electric vehicle

There are a number of different categories for electric vehicles: Neighborhood Electric Vehicles (NEVs) in the United States, Quadricycles in Europe, Tricycles and Cars. The term 'electric car' may refer to any of these. Yet there are important differences between these categories which you need to understand. Here is a brief description of each vehicle category:

Cars

Cars are the mainstream vehicles here. They are designed, built and tested to a heavily prescribed set of standards, covering all aspects of their design and build. They are capable of reasonable performance and, most importantly for our definition, have been subjected to stringent safety regulations in their design and build, incorporating airbags and properly designed and tested crush zones.

With the exception of a few low-volume specialist cars built in Europe, all vehicles classified as cars have to be subjected to crash testing. All cars have had to pass a huge number of checks and tests in order to be certified roadworthy and allowed to be sold.

Virtually all the electric vehicles currently sold by the mainstream car manufacturers such as Nissan, Ford, VW and BMW are technically classed as cars.

The Nissan LEAF is the world's best-selling electric car, with over 200,000 on the roads.

Whilst the legislation for car design has undoubtedly made vehicles extremely safe, there are aspects of this legislation that create barriers for innovation, both from a technical and a commercial point of view.

Technically, there are times when a different approach can work better than the conventional solution. Airbags, for instance, are rightly compulsory in a car capable of driving at high speed, but can actually create worse injuries than they prevent in a low-speed accident. A reinforced crash safety cell around the passenger compartment sounds like a sensible mandatory requirement for all vehicles, until you take into account the fact that these structures create blind spots on the vehicle that increase the likelihood of a low-speed accident in a city.

Commercially speaking, designing and building a car is a hugely expensive business. Naturally, manufacturers want to build vehicles they know will be a commercial success and this inevitably produces conservative designs, with manufacturers being less daring than they would like to be. Cars have become more and more complicated. Creating a new type of vehicle that truly changes the way we think about transportation becomes far more difficult when faced with the restrictions of car design and build legislation.

NEVs

In the United States of America, there is a class of low-speed electric vehicles classified as Neighborhood Electric Vehicles (NEVs). NEVs are lightweight vehicles designed for city use only.

NEVs have become quite popular in some parts of the United States and over 75,000 of them have been registered in the past ten years. They have become popular in retirement communities and are often used as a runabout second vehicle for a household.

In the past, NEVs were restricted to a top speed of 25 mph (40 km/h) and could only be driven on roads with a speed limit of 35 mph or less. This meant that NEVs were extremely restricted in use and were best suited to gated communities, campuses, large parks, estates and central city areas.

Several states have now updated their legislation, creating a new class of 'Medium Speed Vehicle'. In these states, NEVs are now allowed to travel at speeds of up to 35 mph (56 km/h) and can travel on roads with speed limits of up to 45 mph. At the time of writing, Minnesota, Oklahoma, Montana, Washington, Kentucky and Tennessee have passed these laws, with many other states currently considering them.

As a result of these reduced restrictions, NEVs may now be a practical option for Americans who live and work in major cities and predominantly use their cars for short journeys around town.

Small commercial NEVs, mainly trucks and small vans, have become popular with some local authorities, where they are often used for street cleaning, park management and refuse collection in central city areas. The American Army also uses thousands of NEVs as small personnel carriers, for deliveries, and for maintenance duties around military bases.

The Google self-driving car is technically classed as a Neighborhood Electric Vehicle

NEVs are not subject to the same safety regulations as normal cars. Most of them have not undergone crash testing and none of them have airbags. In terms of passenger safety, they are safer than a motorbike but not as safe as a car.

As you look through the list of vehicles on the following pages, you will notice that all NEVs are shown as having a top speed of 25 mph. In most cases, the cars have been electronically restricted to this speed.

In the American states that allow an NEV to travel at 35 mph, the top speed of the vehicle can normally be increased by the dealer.

Low-Speed Vehicles (LSVs)

Canada used to have Low-Speed Vehicles legislation that was similar to the United States' original NEV legislation. However, this was not popular with many states and the legislation was rescinded during 2008, restricting LSVs to gated communities.

As a consequence, no new LSV vehicles are available in Canada. Existing vehicles built before the law was changed can still be used on the roads.

Quadricycles

The European equivalent to NEVs is the Quadricycles category. There are two distinct categories of quadricycles: 'light' quadricycles, with a top speed of 28 mph (45 km/h), which can be driven by 16-year-olds and in many countries can be driven without driving licences, and 'heavy' quadricycles, which are not speed-restricted and in many cases can be driven by owners who only have a motorcycle driving licence.

The Renault Twizy is technically classed as a quadricycle. It's a zippy two-seat car designed for inner city driving.

Quadricycles come in various shapes and sizes. Some are little more than motorcycles on four wheels, whilst others are like compact cars, usually with seating for either two adults or two adults and two children.

Diesel-powered quadricycles have been on sale in Europe since the early 1980s, while electric quadricycles have appeared in the past few years. They are very popular, particularly in France, Italy, the Netherlands and Belgium. In total, there are over 300,000 quadricycles on the roads across Western Europe.

Like their North American equivalent, quadricycles are not subject to the same safety regulations as normal cars and do not have to be subjected to crash testing. Despite this, many manufacturers voluntarily submit their quadricycles to crash testing in order to maintain customer confidence.

Quadricycles and safety

Quadricycles have a very good safety record. In France, Europe's largest quadricycle market, these vehicles are three times less likely to be involved in an accident than a conventional car[11]. Furthermore, accidents with quadricycles are only half as likely to result in serious injuries as accidents with cars.[12]

Why are these classifications important?

Quadricycles and NEVs are, by their very nature, niche products and they will not work for everyone. They are not designed for taking the family on a long journey, for example, but for driving through towns and cities to get to work, for doing the shopping, or for dropping off the kids, they do have benefits. They're small enough to get through city traffic. They can be parked anywhere, squeezing into spaces that no ordinary car could get into, and the running costs are usually very low.

In all likelihood, if you are thinking of getting an electric vehicle today, you are most likely to be buying a car, not a quadricycle or an NEV. Whilst there are a few quadricycles and NEVs available to buy today, they are a very small niche and are largely overlooked by the general public.

However, the real importance of these categories of vehicles lies in what manufacturers can do with them. Inevitably, legislation always struggles to keep up with innovation. It always backs the status quo. Yet electric vehicles are a new technology with the potential to disrupt the status quo and change the way transport works in the future.

The legislation for NEVs and quadricycles allows manufacturers to be more inventive and create new products that otherwise simply could not exist. These products may be simpler and less adaptable than conventional cars, but they provide answers to specific problems, such as how to cut through traffic to reduce journey times, how to find somewhere to park easily, or how to cut the costs of car ownership.

---

[11]The French National Interministerial Road Safety Observatory.

[12]European Quadricycle Manufacturers and Importers Association.

Much of the technology around self-driving cars will first be tried out in NEVs and quadricycles, with compact, self-driving city taxis like the Google driverless car appearing in both North America and Europe.

Chapter summary

- In legislative terms, there are a number of different categories of electric vehicles available today.
- Cars are the most well-known. They are built and tested to a heavily prescribed set of standards.
- Neighborhood Electric Vehicles (NEVs) in the United States are a low-speed alternative to cars that are designed for inner city driving.
- The European equivalent of NEVs are quadricycles. These are faster and have fewer restrictions on use than NEVs.
- The benefits of these different classifications are that they allow manufacturers to be more creative and innovative with new vehicle designs.
- Many new vehicle technologies, such as self-driving cars, are likely to be pioneered with NEVs and quadricycles, rather than with more conventional cars.

Electric Cars You Can Buy Today

I wrote the first version of this book in 2010, at a time when there were very few electric cars on the road. Most of the cars from mainstream manufacturers were prototypes and most of the vehicles on sale came from small independent car makers.

Today, it is a very different story. A few of the independent manufacturers remain, but the majority of the cars come from mainstream manufacturers. Some of them have built dedicated electric car models, whilst others have produced electric-drive variants of existing models.

Not all of the cars listed in this chapter are available everywhere. Some are only sold in Europe, a few are only available in one or two states in the US, and some are discontinued but available second hand.

I've tried to give an objective view of each car. I've driven most of the cars that are available and spoken to other owners. I've also given an idea of what they cost to run and how far you can expect to drive, and provided a brief summary that explains, in my view, what is good and what is not so good about the car.

Of course, the summary and description is only my view, and my opinions about what makes one car better than another won't necessarily match with yours. If you like the look of a car, go and test one for yourself and see what you think.

Understanding the specifications

For each car, I've included a table that shows basic information about each vehicle:

Purchase Price (US)	from $xxxxx	NEDC Range (miles)	xxx miles
Purchase Price (UK)	from £xxxxx	NEDC Range (km)	xxx km
Rapid charge	Yes (CCS)	EPA range (miles)	xxx miles
Standard charge time	xx hours		

Purchase Price

The purchase price is shown for the United States and the UK.

The US price *excludes* the current $7,500 federal tax rebate, plus any other discounts offered at state level. For example, California currently offers a further $2,500 tax rebate for new electric car purchases.

This means that the US price on the specification looks much higher than the eventual price you will end up paying.

In the UK, the government offers a grant of up to £5,000, which is applied as a discount from the retail price. The UK purchase price is shown with the discount *already applied*, as this is the price you will pay for your car in the showroom.

Rapid Charge

If the vehicle can be fast-charged from a high-power rapid charging-station, I include details of what rapid charging-stations it can use: CHAdeMO, CCS or AC. See the chapter on Electric Car Charging Networks, starting on page 67, for more information about rapid-charging options.

Standard Charge Time

The standard charge time is the time it takes to charge the car up at home using a dedicated electric car charging-point.

NEDC Range (miles and km)

The European standard range measurements (NEDC) are notoriously optimistic and bear little resemblance to a real-world range. If you are buying a car in Europe, these are the figures the manufacturers will quote to you.

EPA Range (miles)

The US standard range measurements (EPA) are far closer to the real-world range you will get from an electric car. If you are buying a car in the United States, these are the figures the manufacturers will quote to you.

BMW i3

Purchase price (US)	from $41,350	NEDC range (miles)	118 miles
Purchase price (UK)	from £25,680	NEDC range (km)	190 km
Rapid charge	Optional (CCS)	EPA range (miles)	72 miles
Standard charge time	3½ hours		

Without doubt, the BMW i3 is one of the most exciting electric cars available today. Fast and entertaining to drive, the i3 is built with the quality you would expect from BMW, with a light and airy cabin and eye-catching styling.

It is the most advanced electric car available today. Unlike other manufacturers, BMW have designed the i3 from the ground up as an electric car. Built from advanced carbon-fibre composite materials, the result is a strong, lightweight car that gives better real-world economy than most other electric cars. It's not without its faults: access to the rear seats is not the easiest and the exterior styling is love-it-or-hate-it, but it has won over a lot of confirmed petrol heads. Notable owners include James May, one of the three former presenters of *Top Gear* in the United Kingdom, and BMW dealers are reporting strong sales, both to existing BMW owners and to new customers who have never owned a BMW before.

Compared to conventional cars, it is pricey as a city car, but not expensive by BMW standards as a premium small hatchback. Is it good value for money? Running costs are very low and depreciation is the lowest in its class, so if you are travelling higher than average mileages, an i3 offers good value for money in a stylish, fun-to-drive car.

BMW do two different versions of the i3: an electric-only car and an electric car with a small two-cylinder motorbike engine that acts as a generator to keep the batteries charged up for longer journeys. As a 'hybrid', the car is not particularly economical: expect around 40 mpg if driving on fuel alone, but as a way of extending the range of the car for occasional longer journeys instead of recharging at rapid charging-points, it is very effective.

If you are buying the all-electric version, it is worth noting that the rapid charging facility is an option on the i3. This means that if you want to charge your car in around 30 minutes, you will need to specify the DC rapid-charge option.

It's good for:

- ✓ Performance, handling, fun.
- ✓ Stylish, head-turning looks
- ✓ Beautifully finished interior
- ✓ Better real-world range than a Nissan LEAF or VW Golf

It's not good for:

- ✗ Awkward rear entry and small luggage area
- ✗ Not particularly economical as a hybrid
- ✗ DC rapid-charge is a costly optional extra

Chevrolet Bolt

Purchase price (US)	from $37,500	NEDC range (miles)	N/A
Purchase price (UK)	N/A	NEDC range (km)	N/A
Fast charge	Yes (CCS)	EPA range rating	N/A
Standard charge time	9 hours	Claimed range	200 miles

Chevrolet's second all-electric car was shown for the first time as a concept vehicle in 2014 and reaches production later in 2016. Initially only to be available in the United States, the Bolt is Chevrolet's answer to the Nissan LEAF. Reviewers who have driven pre-launch models of the car have been impressed, particularly with its driveability and handling.

At the time of writing official EPA range testing is yet to be complete, but Chevrolet are claiming a real world range of over 200 miles between charges, with a full charge taking around nine hours.

Chevrolet Spark EV

Purchase price (US)	from $25,995	NEDC range (miles)	N/A
Purchase price (UK)	N/A	NEDC range (km)	N/A
Fast charge	Yes (CCS)	EPA range rating	82 miles
Standard charge time	7 hours		

The Chevrolet Spark is a small, four-door, four-seat city car that has been on sale in Europe and Asia as a conventional engined car for around four years. Designed for city driving, the car is compact, manoeuvrable and easy to drive.

Performance in the Spark EV is surprisingly quick: it easily outperforms every other Spark, with a 0–60 mph (0–100 km/h) time of around 8 seconds. Range is good, comfortably better than most of its rivals.

The Chevrolet Spark EV is currently only available in a handful of states in the US, but is often available with fantastic finance deals, often as low as $139 a month on lease.

It's good for:

- ✓ Rapid performance
- ✓ Reasonably priced
- ✓ A good around-town car

It's not good for:

- ✗ Uncomfortable for taller occupants
- ✗ Dark, uninspiring interiors

Citroen C-Zero / Mitsubishi i-MiEV / Peugeot iOn

Purchase price (US)	from $22,995	NEDC range (miles)	93 miles
Purchase price (UK)	from £11,995	NEDC range (km)	150 km
Fast charge	Yes (CHAdeMO)	EPA range rating	62 miles
Standard charge time	6 hours		

The Citroen C-Zero, the Mitsubishi i-MiEV and the Peugeot iOn are the same car, built in the same factory, with minor trim and configuration changes to differentiate them. Only the Mitsubishi version is available in North America.

The cabins are light and airy, the high-up seating position gives great visibility, they are comfortable for taller drivers, they perform well and they are fun to drive.

In the United States, the i-MiEV is one of the cheapest electric cars available, with a federal tax credit bringing the price down to $15,495, comparable to the purchase price of an equivalent gasoline car. In Europe, Citroen have slashed the price of the car, bringing it down to under £12,000, offering great value for money.

It's good for:

- ✓ Fun and easy to drive
- ✓ Great manoeuvrability
- ✓ Can be excellent value for money

It's not good for:

- ✗ Spartan interior
- ✗ New list prices in Europe are high, so shop around

FIAT 500e

Purchase price (US)	from $31,800	NEDC range (miles)	N/A
Purchase price (UK)	N/A	NEDC range (km)	N/A
Fast charge	No	EPA range rating	87 miles
Standard charge time	4 hours		

Europeans adore small cars and the car they love the most is the FIAT 500. With stylish looks and dynamics that encourage sporty driving, it's a popular choice for city driving.

The electric FIAT 500e was introduced in California in 2013. Compared to the conventional combustion engine version, the electric FIAT sacrifices luggage space and rear leg room, making it only really suitable for two adults and two young children. In return, you get a powerful electric motor and excellent driving dynamics that have impressed the motoring journalists. Top Gear claim that the 500e 'is one of the all-time best city cars ever made'. Unfortunately, only Californians get to enjoy the 500e. FIAT has no plans to introduce the car elsewhere.

It's good for:

- ✓ Stylish, good looking
- ✓ Huge fun to drive
- ✓ Good range

It's not good for:

- ✗ Cramped rear seats
- ✗ Limited luggage space
- ✗ No fast-charge capability

Ford Focus Electric

Purchase price (US)	from $29,995	NEDC range (miles)	100 miles
Purchase price (UK)	from £28,580	NEDC range (km)	160 km
Fast charge	Yes (CCS)	EPA range rating	76 miles
Standard charge time	3-4 hours		

Unsurprisingly, the Ford Focus Electric is quite simply an electric version of the Focus. Ford have managed to keep the driving dynamics and agility from the standard cars and the only main difference is a much smaller luggage space, where the batteries intrude, and a digital dashboard to help drivers drive more efficiently.

The Ford Focus is often regarded as the best drivers' car in its segment, and the Focus Electric is no exception. It is more entertaining to drive than the Nissan LEAF or Renault Fluence ZE, for example, although most drivers who want an entertaining car will probably prefer the dynamics of the smaller BMW i3.

The Focus Electric is only available through a very few dealers and this, combined with its high price, has made it a poor seller.

It's good for:

- ✓ A good drivers car
- ✓ Smooth power delivery
- ✓ Well liked by owners

It's not good for:

- ✗ Limited luggage space
- ✗ Expensive in Europe
- ✗ Very few dealers

KIA Soul EV

Purchase price (US)	from $31,950	NEDC range (miles)	120 miles
Purchase price (UK)	from £24,450	NEDC range (km)	192 km
Fast charge	Yes (CHAdeMO)	EPA range rating	93 miles
Standard charge time	5 hours		

The KIA Soul is a compact cross-over vehicle that has become popular in Europe. The electric version has an impressive range, a competitive price and comes with KIA's standard seven-year warranty. It will appeal to families who will appreciate the car's versatility and spaciousness, despite its compact overall dimensions.

The KIA has one of the largest battery packs around, giving it a useful extra range over its competitors. Ride is soft and comfortable, but at the expensive of handling.

The KIA Soul EV is available across Europe, but will initially be only available in California, Oregon, New York, New Jersey and Maryland in the United States.

It's good for:

- ✓ Spacious, practical family car
- ✓ Longer range than its competitors
- ✓ Competitively priced

It's not good for:

- ✗ Not a driver's car
- ✗ Limited availability in the US

Mahindra e2o

Purchase price (US)	N/A	NEDC range (miles)	79
Purchase price (UK)	from £12,999	NEDC Range (km)	127
Fast charge	Yes (CHAdeMO)	EPA range rating	N/A
Standard charge time	9 hours		

Mahindra has more experience than most with electric vehicles, having purchased electric car specialist REVA, who launched their first electric car in 2001. This gained a cult following in London where it was known as the G-Wiz, and for many years was the best-selling electric car in the world.

Their latest offering, the e2o, has been on sale for a while in India, but has been extensively updated for its European launch. It's a compact, three-door, four-seat hatchback with stylish looks and a smart interior that offers a surprisingly large amount of room and belies the compact size of the car.

Unashamedly a city car, the e2o is one of the most efficient electric cars available today and has top environmental credentials, thanks to its new state of the art manufacturing site and a design that focuses on efficiency. It boasts a tiny turning circle and good all-round vision, thanks to its high up seating position and large glass area. It has a small battery pack, but gives a reasonable range thanks to its light weight and compact size.

It's good for:

- ✓ Eye-catching looks, smart interior
- ✓ Good interior space
- ✓ One of the most efficient electric cars available today

It's not good for:

- ✗ Awkward rear entry
- ✗ Tiny luggage space
- ✗ Modest performance

Mercedes B-Class Electric Drive

Purchase price (US)	from $41,450	NEDC range (miles)	124 miles
Purchase price (UK)	from £32,275	NEDC range (km)	200 km
Fast charge	No	EPA range rating	87 miles
Standard charge time	3½ hours		

The Mercedes B-Class Electric Drive brings a premium brand to the family-sized electric car segment, currently dominated by the Nissan LEAF.

The B-Class Electric Drive is certainly very quick; 0–60 mph (0–100 km/h) takes just 7.9 seconds and the official 87-mile range is a useful improvement over the competition. Mercedes also offers a 14-mile range upgrade, which allows owners to overcharge the batteries for occasions when they are planning to drive longer distances.

Unfortunately, the B-Class Electric Drive does not include rapid charging capabilities, so true long-distance driving is not a practical proposition.

It's good for:

- ✓ Spacious and practical
- ✓ Reasonable range
- ✓ Rapid performance
- ✓ Premium brand

It's not good for:

- ✗ No rapid charge capability limits usefulness for longer journeys

Nissan e-NV200

Purchase price (US)	from $23,125	NEDC range (miles)	106 miles
Purchase price (UK)	from £13,393	NEDC range (km)	170 km
Fast charge	Yes (CHAdeMO)	EPA range rating	TBA
Standard charge time	3½ hours		

The Nissan e-NV200 is an electric van that is also available as a five-seat passenger car. Based on the same battery pack and motor as the Nissan LEAF, the all-electric e-NV200 has a 703kg (1550 lb) payload, or seating for five or seven, depending on the version required. A taxi version arrived in 2015.

On sale in Europe since mid-2014, and in the US since early 2015, Nissan have kept the cost low by allowing customers to lease the batteries instead of buying them outright. Battery lease costs depend on mileage, but start from £61 per month (US prices yet to be announced). If customers prefer, they can buy the batteries with the vehicle at an additional cost of £3,169 (in the UK).

Nissan have four models available. The bottom-of-the-range *Acenta* takes longer to charge (6 hours instead of 3½) and does not have a CHAdeMO connector for high-power rapid charging.

It's good for:	*It's not good for:*
✓ Big, spacious and practical ✓ Competitively priced against diesel alternative	✗ Rapid charging not included in base model ✗ Confusing sat nav touch screen

Nissan LEAF

Purchase price (US)	from $29,010	NEDC range (miles)	155 miles
Purchase price (UK)	from £16,490	NEDC range (km)	300 km
Fast charge	Yes (CHAdeMO)	EPA range rating	107 miles
Standard charge time	3½ hours		

The Nissan LEAF was first introduced at the end of 2010 and along with the Mitsubishi i-MiEV, was one of the first electric cars to come from a mainstream manufacturer.

The car received critical acclaim in the motoring press, winning 2011 European Car of the Year and 2011 World Car of the Year awards. Nissan upgraded the car in 2013, with reduced prices, improved range and faster home charging – down to 3½ hours from the previous 6 hours.

For 2016, Nissan launched a new, larger battery model, increasing the range significantly, making it far more useful for longer journeys on faster roads.

The LEAF is still the benchmark by which other electric cars are judged. It's smooth, quiet and spacious, with suspension that is tuned for comfort. It is the best-selling electric car in the world.

If you want an electric car but can't afford new, a used Nissan LEAF is well worth a look. Early cars are available at extremely good prices and offer excellent value. Long

term battery performance has been proven to be good, with many owners now having driven over 100,000 miles without fault.

Since 2013, Nissan has offered the option of leasing the batteries rather than buying them outright. This reduces the list price of the car so that it is comparable in price to a conventional car with an internal combustion engine.

Like the e-NV200, the bottom-of-the-range Nissan LEAF has a slower standard charge time (6 hours) and does not have a CHAdeMO connector for high-power rapid charging.

It's good for:	It's not good for:
✓ Spacious, practical, good family car ✓ Good performance ✓ Distinctive looks ✓ Good choice of used cars	✗ Soft ride won't suit some drivers ✗ Confusing sat nav touch screen ✗ Rapid charging not included in base model

Renault Kangoo ZE

Purchase price (US)	N/A	NEDC range (miles)	106 miles
Purchase price (UK)	from £16,313	NEDC range (km)	170 km
Fast charge	No	EPA range rating	N/A
Standard charge time	8 hours	Makers' claimed range	78 miles

There are two versions of Kangoo ZE: a short wheel base van, or a long wheel base model, available as a van or a crew cab estate car. The list price excludes the battery, which has to be rented on a monthly basis.

It may be a van, but it is surprisingly enjoyable to drive. Unpretentious and easy to drive, it makes a great city delivery van. The lack of a rapid-charging option is a drawback and it now looks expensive when compared to the bigger and more powerful Nissan e-NV200.

It's good for:

✓ A practical van or estate car
✓ Rewarding to drive

It's not good for:

✗ Lack of rapid-charging option

Renault Twizy

Purchase price (US)	N/A	NEDC range (miles)	62 miles
Purchase price (UK)	from £6,895	NEDC range (km)	100 km
Fast charge	No	EPA range rating	N/A
Standard charge time	3½ hours		

The Renault Twizy is a funky-styled city car, classified as a quadricycle in Europe. Half car, half quad bike, it is one of the cheapest electric vehicles you can buy today.

It's huge fun to drive. Seating is strictly two people – one in the front and one in the back. Doors are optional and even if you have them, they aren't fully weather-proof. But that is all part of the charm of this fun, quirky vehicle. It's designed for inner-city driving and its tiny size means it can be parked anywhere. Charging is through a standard domestic socket and takes 3½ hours from flat to full.

Twizy is a great choice for younger drivers. Great visibility, airbags and a front crumple zone like a full-sized car. Top speed is 50 mph (80 km/h), so it discourages racing. And its low purchase price and running costs makes it cheap to use.

Twizy has not made it to the US. However, you can hire one by the hour by Scoot Networks in San Francisco, as a fun and convenient way to buzz around the city.

It's good for:

- ✓ Funky styling
- ✓ A blast to drive
- ✓ Cheap to buy

It's not good for:

- ✗ Practicality? What practicality?
- ✗ Even with doors, it isn't weather-proof

Renault Zoe

Purchase price (US)	N/A	NEDC range (miles)	130 miles
Purchase price (UK)	from £13,444	NEDC range (km)	210 km
Fast charge	Yes (AC)	EPA range rating	N/A
Standard charge time	3½ hours		

The Renault Zoe is a capable super-mini, with seating for five adults, a bright interior and one of the most user-friendly dashboards on any car today.

It has a clever charging system that allows you to charge up from any AC power supply: anything from a domestic socket, which takes around 7 hours, to a three-phase high-current supply that can charge the car up in around half an hour. This gives a huge amount of flexibility on when and where to charge up, particularly as Renault has invested in installing AC rapid-charging at motorway services in the UK.

Performance is reasonable, although it is no sports car. Range is good, too, noticeably better than the Nissan LEAF, and thanks to a clever heating system, range does not drop significantly when using heating or air conditioning. It's a good choice for frequent longer journeys.

Batteries are leased separately to buying the car. This keeps the purchase price down, but does mean an ongoing monthly cost, which is unpopular with some buyers. Renault is now allowing buyers to purchase the car with batteries if they prefer.

It's good for:

- ✓ Good range
- ✓ Flexible charging solution
- ✓ Good interior ergonomics

It's not good for:

- ✗ Battery leasing is unpopular with many buyers

Tesla Model S

Purchase price (US)	from $71,100	NEDC range (miles)	340 miles
Purchase price (UK)	from £52,335	NEDC range (km)	547 km
Fast charge	Yes (Supercharger)	EPA range rating	265 miles
Standard charge time	4½ hours		

Ten years ago, the world had never heard of a car company called Tesla. Today they are the fastest growing premium car brand in the world, exclusively building electric cars.

Their Model S is one of the most desirable luxury cars around. In the US, they regularly outsell the BMW 7-series, Mercedes S-Class, Audi A8 and Jaguar XJ-8 combined. They are also becoming a common sight on roads in the UK. In 2015, Tesla Model S sales overtook sales of the far cheaper Nissan LEAF.

It's easy to see why. The car oozes luxury appeal, with great looks and a plush interior complete with 17-inch touch screen which eliminates a lot of dashboard clutter and gives an up-market, high-tech look. Its performance figures are more akin to an Aston Martin or a Ferrari than a Mercedes. Even the bottom-of-the-range model has a 380 bhp motor, whilst the top-of-the-range four-wheel drive, dual motor model has over 700 bhp and is capable of a 0–60 mph time of 2.8 seconds.

Since its original launch in 2013, Tesla have constantly updated the car, ensuring it has the latest technology, including its 'autopilot' technology with automatic steering, lane changing and parking technology. Many of these updates are provided free of charge to existing owners, via wireless software updates.

The Model S can travel hundreds of miles without needing to be charged up, and it is backed up by Tesla's own national network of 'supercharger' charging-stations that can give a 170-mile range boost in as little as 30 minutes, all for free.

It's good for:

- ✓ Latest technology
- ✓ Long distance cruising
- ✓ Terrific performance
- ✓ Luxury driving

It's not good for:

- ✗ It's a big car by European standards. If your drive includes lots of small country lanes, look elsewhere!

Tesla Model X

Purchase price (US)	$80,000	NEDC range (miles)	310 miles
Purchase price (UK)	£65,000	NEDC range (km)	498 km
Fast charge	Yes (Supercharger)	EPA range rating	265 miles
Standard charge time	6 hours		

A luxury, all-electric, seven-seat, four-wheel drive SUV with a 0–60 mph time in less than 5 seconds and an official range of 265 miles? That's the new Tesla Model X. The car went on sale in the US in 2015, with European sales starting this year.

Tesla have already turned the luxury car market on its head with the Model S, and the Model X looks certain to cement its position at the top of the premium car market. The car has very distinctive looks, including 'falcon wing' rear doors allowing for easy access to the two rear rows of seats. The wood-and-leather dashboard is dominated by a huge 17-inch touch screen display, providing all the information and entertainment systems and eliminating clutter from the dashboard.

It's good for:

- ✓ Stylish luxury, inside and out
- ✓ Terrific performance
- ✓ Four wheel drive
- ✓ Seven seats

It's not good for:

- ✗ By European standards, it's huge! It's more at home on Californian highways than country lanes in Europe.

Tesla Model 3

Purchase price (US)	$35,000	NEDC range (miles)	TBA
Purchase price (UK)	TBA	NEDC range (km)	TBA
Fast charge	Yes (Supercharger)	EPA range rating	215 miles
Standard charge time	6 hours		

You won't be able to own one in 2016, but the order books are now open for Tesla's latest creation, the Model 3. Described by many as Tesla's BMW 3-series beater, the Model III brings Tesla's ethos of high performance, luxury and technology to the mainstream market. Undoubtedly the most important new car for Tesla, the car could be set for making huge inroads into the small executive car market currently dominated by BMW, Audi, Mercedes and Lexus, when it finally arrives at the end of 2017.

When Tesla launched the Model 3 at the end of March, there were queues of hundreds of people at their stores, waiting to put down their deposits on a new car. In total, over 115,000 people placed an order – and paid a $1,000 deposit – in the first few hours of the car being offered on sale, with over quarter of a million cars ordered in the first few days. Tesla is confidently predicting production of 500,000 cars a year.

As you would expect from Tesla, performance and range is good. 0–60 in under six seconds, autopilot safety features and a claimed range of 215 miles puts this car into a different league than other electric cars in its segment and puts Tesla firmly into the

mainstream. Pricing is good, too. The US price of $35,000 is before tax incentives, meaning most Americans will end up paying around $27,500 for their car. There is currently no news about UK or European pricing yet, but expect it to be competitive.

Tesla claim class-leading interior space for the Model 3, whilst its glass roof makes the interior look light and airy. Safety is not compromised either: with five star safety ratings, Tesla claim that the Model 3 is the safest car in its class.

It's good for:	It's not good for:
✓ Competitively priced car from a luxury car maker ✓ Stylish looks ✓ Spacious interior ✓ Great performance ✓ Great range	✗ The competition – BMW, Audi and Mercedes have already lost their premium segment customers to Tesla, now they risk losing their mainstream customers too.

Tesla Roadster

Purchase price (US)	N/A	NEDC range (miles)	245 miles
Purchase price (UK)	N/A	NEDC range (km)	392 km
Fast charge	Yes (Supercharger)	EPA range rating	244 miles
Standard charge time	8 hours		

It's no longer available new, but the Tesla Roadster is the car that shot Tesla to fame. A terrific-looking out-and-out sports car with sub 4 second 0–60mph acceleration and great handling, the car was an instant hit with customers and journalists alike. Very few other cars attract as much attention as the Tesla Roadster when out and about. Park it next to a Ferrari, and it's the Tesla that gets the majority of the attention from passers-by.

Built between 2008 and 2012, around 2,500 Roadsters were built. Used cars appear regularly for sale, mainly sold through Tesla. Tesla still offers new upgrades for the car. The most recent is a battery upgrade to give a range of 400 miles, making it the longest-range electric car available today.

It's good for:

- ✓ Out and out sports car
- ✓ Head-turning looks
- ✓ Terrific performance
- ✓ Impressive range

It's not good for:

- ✗ Build quality is not up to the same standard as a Porsche
- ✗ Cramped interior, not the easiest car to get into or out of

VW e-Golf

Purchase price (US)	from $36,300	NEDC range (miles)	118 miles
Purchase price (UK)	from £26,650	NEDC range (km)	190 km
Fast charge	Yes (CCS)	EPA range rating	85 miles
Standard charge time	4 hours		

Comparable in size to the Nissan LEAF, the VW e-Golf has a solid, Germanic feel and a more conventional interior.

Three driving modes – Normal, Eco and Eco Plus – allow drivers to choose between power and economy. Normal provides good performance, whilst Eco Plus is best used for city and urban driving.

It's good for:

- ✓ Solid Germanic build quality
- ✓ Premium feel to the interior
- ✓ Just like every other Golf

It's not good for:

- ✗ Considerably more expensive than a Nissan LEAF.
- ✗ Just like every other Golf.

VW e-Up!

Purchase price (US)	N/A	NEDC range (miles)	93 miles
Purchase price (UK)	from £19,250	NEDC range (km)	150 km
Fast charge	Optional (CCS)	EPA range rating	N/A
Standard charge time	6 hours		

VW's tiny city car, the Up! also comes as the *e-Up!*, an all-electric car. It is a fun car to drive around town, where its sharp steering and instant torque make it an entertaining drive.

It's good for:

- ✓ Fun car to drive in town
- ✓ Good ride and handling

It's not good for:

- ✗ Rapid charging is an optional extra

Me and My Electric Car

Electric car owners come from all walks of life. Some of them own their cars because of the environmental benefits, others because they want to drive something different. Everyone has their own reasons.

With the exception of this introduction, this entire chapter is written by other people. These people already use an electric car as their day-to-day vehicle. This is their story.

Neil Butcher – Mitsubishi i-MiEV — United Kingdom

We have been driving a Mitsubishi i-MiEV for nine months now and it has not let us down once.

After the first week of novelty, driving an i-MiEV is just like driving any other car. The lack of gear changes makes the ride very smooth and the driving experience very relaxed. The car has four adult seats and enough room for the weekly shop.

The car is fine for all types of road, including motorways, although sharp cornering at speed is not the best. The car is very quiet at slow speeds, but wind noise increases at higher speeds and there is a whine from the electric motor at 70 mph (112 km/h). The Alpine in-car entertainment system provides good quality sound.

The car is used for my commuting to work, 20 miles (32 km) each way, plus all the local shopping and social journeys. My wife also has a car and we swap cars if I need to make a long business journey. My longest single journey has been 55 miles to Bedford, down the M1 at 65 mph (104km/h), and I still had 4 miles range left. It does not sound a lot but I was always confident that I would make it.

I tend to drive in boost mode most of the time, which has enhanced regenerative braking to recover energy back to the batteries. When you take your foot off the accelerator, the mode slows the vehicle down more quickly than a conventional car, so you need to be a little considerate to any cars close behind: the brake lights do not come on until you actually press the brake pedal. I have tended to change my driving style and anticipate the need to brake to take full advantage of the regeneration, and because the brakes by themselves are not the sharpest on the road.

I am sure that I could get 80 miles (128 km) range from the vehicle with urban driving, but my commute journey has some dual carriageway and the range, like that of any other vehicle, drops at high speed. I have achieved 50-55 miles range at 70 mph (80-90 km range at 112 km/h).

We had the vehicle during the very cold January 2010 weather. It was noticeable how much the range dropped during this period. This was in part due to the use of lights, heater, and windscreen wipers, but was also due to the battery not accepting as much charge in cold conditions. In the very worst weather I was still achieving 50 miles (80 km) range on my commute, compared to around 70-75 miles (112-120 km) in summer.

Other than the reduced range, the main difference to conventional cars is of course fuel. I am lucky in that I have a charging-point at home and work, so I have some flexibility. I usually plug in at both home and work, by habit rather than by necessity – it only takes 20 seconds. I find it far easier, quicker, cheaper and cleaner to do this than to stop off at a petrol station once per week!

We have used the public charging-points a few times, but rarely because we needed them to complete the journey. They are like insurance – there for peace of mind, but rarely needed.

In summary, we have been very pleased with the i-MiEV and would be happy to keep an electric vehicle in the family when we return it at the end of its twelve-month lease.

Peter Berg – Nissan LEAF California, USA

I have always been interested in green technology. In 2002, I bought my first new car ever, a Toyota Prius hybrid. It was a beautiful cobalt blue and was the most efficient car I had ever driven. At the time, it seemed that the Prius was the most advanced and efficient vehicle that was available.

I was not aware of any electric vehicle options until I read an article about the new Aptera EV that was being developed. My interest peaked, I paid the deposit and got on their waiting list. I was bitten by the electric vehicle bug and I wanted to be a part of the revolution I knew was coming.

The turning point for me was when Nissan announced their new LEAF electric car. They hosted an event here in Los Angeles and invited those interested in being early adopters. At the event, I saw a prototype of the car, learned about how their car would work and saw their battery technology. It was then that I knew I wanted an electric vehicle. I got on Nissan's list and paid the reservation fee right away.

In December of 2010, after eight months on the waiting list, I received an email telling me that our car had been produced in Japan and was on its way to California. What a wonderful early Christmas present! We got a call from our dealer a few weeks later and picked up our all-electric car in January 2011.

Our dealer was sixty miles away in Fontana, California. Danny, the sales rep, called on a Tuesday to let us know our Cayenne Red LEAF was ready. Did we want to pick it up tonight or wait until the weekend to make the drive out there? My choice was obvious. We went to the dealer after work and arrived a little after 9 pm. When we got there, the staff were as thrilled as we were, despite the late hour. The manager stayed late to make sure she shook our hands, and photos were taken. We were some of the first people in the United States to get this exciting new car.

The car comes with great features: a built-in GPS navigation system, Bluetooth technology to connect to my phone, XM satellite radio, live traffic information, a USB port to play music with, and more. Nissan did a great job with the engineering, and the build of the car is sleek and modern. It has definitely gotten some attention, both on the road and off.

Normally I only drive about 30 miles per day so the car works great for my daily work commute. We have been able to take it on a few longer trips as well. We went to visit my wife's parents in Orange County, and it was an easy trip there and back. We also took a trip up into the hills to go for a weekend hike, and it handled the hills and curves with ease.

Keeping the car charged has been equally trouble-free. After one of those longer trips, the car can be charged in about six to seven hours, but with my short commute I can get away with plugging in only two or three times a week. We might have been able to use only our main 120 volt power source to charge the car, but I thought the faster 240 volt option would be good when we need a faster charge. I found a company that made an EVSE (electric vehicle supply equipment, or charging dock) that worked with the LEAF and I installed it myself. With the solar panels on our roof charging the car from the sun, we now are able to lower our power bill *and* our carbon footprint.

Owning the LEAF has made me think a little bit differently about driving now. I have to do a small amount of planning to make sure we will have enough charge for our trip and to decide which vehicle to take (we still have our Prius). While some people say that doing that planning is too much for them or too inconvenient, I have found that it is not a big deal. It is certainly worth the advantages that come with the car: no more trips to the gas station, much less maintenance with the car, a smoother and quieter ride, lower operating costs with the car, the cutting-edge technology that comes with the car, and, most importantly, less pollution going into our air.

There have been a few issues with the car, but nothing major. I did have to get used to the charging settings. You can set up timers to delay the charging until the evening but I had to learn how the charging timers worked. The first night, my car did not start charging since it was after the start time on the timer. I used the override switch to solve the problem for the first night. I modified the timer and now it is not a problem.

Despite these few small issues, we love the car. It is easily the most fun car I have had. While I wanted a fun car, our main reason for getting the car was to make the air we breathe a little cleaner. Here in Los Angeles our air quality is not good. It is often hazy and unhealthy to breathe. It is obvious that the air quality is a result of man's activities, so it is only fair for us to step up now to do more to reduce the pollution going into the air and water. The secondary reasons to get the car were the lower operational and maintenance costs, the rebates that made the cost of the car very reasonable, and the smoother ride that the car gives us.

I think electric cars, and the LEAF in particular, are a great option for many people. Anyone interested in getting an electric vehicle should at least test one out. If you see how far you really drive in a day you might be surprised how easy it would be to start driving an electric vehicle. The car keeps you informed about how far you can go, so you do not have to worry so much. Range in an electric vehicle will be no more a concern than in a traditional gas car once more charging-stations are installed in public areas. Just do not go out on long cross-country trips: rent a car for those instead of putting the wear and tear on your own car.

Having a clean and green electric car is a great choice, and I hope a lot more people get rid of the gas-guzzlers and start driving one of these fun and efficient electric cars. They really are a great way to go.

Paul Edgington – Renault Fluence ZE   United Kingdom

Living with an EV is, I suspect, rather akin to having a baby. The anticipation is great, the euphoria on arrival is breath-taking, the reality is rather different. There is nothing to compare with the silent and effortless wafting along that awaits the purchaser. The performance figures may not be in the sports car class but the instant surge of power as the accelerator is floored is truly impressive. The sheer delight of passing petrol stations without reaching for one's wallet is reassuring.

In the cold light of day all is not quite so rosy. However the instruments are configured, the visual evidence of a rapid demise of charge could make you think there's a hole in the tank and reach for a map. A map? Indeed, searching for a charge-point becomes a never-ending quest for energy that a *petrol head* would never imagine in his worst nightmare. Where can I charge? Will I make it? Will the car fit it? Will it work? Will I have the necessary permission? Will that bod in his beaten-up Escort have snaffled the parking?

Rapidly one rethinks how an EV is to be used. As a commuter within range it is great, for a two-car family with alternatives for longer runs it is practical. For the truly adventurous the joy of achieving a long journey complete with en route charges is akin to scaling Everest. For the school run it's a great alternative to the missus's 4x4. And if all else fails, it's a great talking point in the pub.

After a while one settles into a routine, accepting the car's foibles, and relaxing. Then something happens. A warning light, a strange noise, a decision made by the car without consulting you not to charge. You reach for the phone and you discover that, already, you know more than the Customer Services manager on the end of the line. You break into a sweat as you try vainly to explain your position to someone whose eye is fixed on an unhelpful computer screen script. "You must remember, this is new technology, we're still feeling our way…" mutters the operative, and thoughts of murder or even suicide cross your mind. The car is scraped up and carted off to the dealer where it sits, all forlorn, awaiting inspiration.

It took 100 years for an ICE car to reach commodity status. Diesels became generally acceptable rather faster. Why should it be any different with an EV? With the rapid proliferation of new technological devices such as iPads, iPods, smartphones and the like we have come to expect instant gratification and a new improved model just a couple of years down the line, when we can discard the last new model like a piece of flotsam. EVs have appeared without sufficient thought, a mad race by manufacturers to beat the opposition, and as a consequence early adopters are paying the price. Things may settle down, unless a newer technology rears its head in the meantime to confuse the issue. Meanwhile, enjoy the ride… I bet you like the roller coaster, too!

Zarla Harriman – G-Wiz *i* United Kingdom

The G-Wiz seemed the obvious choice for our family. I am always very keen to embrace anything that saves our Earth's resources, and it seemed like a good bargain when it came to running costs too. My husband is a bit of a techie, and loved the idea of the G-Wiz's technology and simplicity. Our son, however, thought it was OK as long as he could have a go at driving it!

Our plan was to reduce our running costs, tax and insurance bills. We looked into electric cars, and the only affordable model seemed to be the G-Wiz.

Its range and power seemed to be totally adequate for our needs, as I work in Leicester city and do about 20–25 miles per day. The range of the G-Wiz seemed to give room for those emergency detours over and above the planned ones, such as the mad dash to the supermarket on the way home for cat litter!

I love the 'cutesy' design of it, and the quirkiness of the interior. I love being the focus of attention as I drive about; people even stopping to take photos with their phones as we drive by! I *love* driving past petrol stations, telling people how much it costs per mile, and giving people rides in it. The parking is fantastic, using the tiniest spaces. There are *always* spaces free at the Leicester city car parks with electric car charge-points.

There are not many things I dislike about this great little car. It plugs into my shed and is always full next morning; it is easy to drive and fun to own.

The heater, however, does not do what it says on the tin... and the draughts coming through the doors are positively freezing in winter! But in the snow, never a slip or slide – much better than any car I have ever owned.

I would buy another one, and what more of a recommendation could you have? However, next time I would like more convenient servicing and the lithium batteries!

'ECarFan' – Tesla Model S   California

Electric cars had always intrigued me, but their range was too limiting and generally their styling was either unimaginative or deliberately quirky. Tesla changed all that with the Roadster in 2008 and then the Model S in 2012, cars that not only had a useful range but were intelligently engineered and a joy to drive. As of this writing, there is still nothing on the market that they can be directly compared to.

My work requires that I be available on a moment's notice to drive up to 200 miles roundtrip, and when I'm on call at work, at any time of day. With Model S (I have the 85 kW version) that is no problem, and in addition the growing network of Tesla Superchargers means that the Model S is a fantastic long-distance road trip vehicle. I have driven it over 400 miles in a single day, and with the usual stops for food and just to stretch my legs the total trip time was only minutes longer than in a conventional car.

My initial experience of living with the car (as opposed to a few brief test drives) confirmed that I had made the right decision. Never having to go to the gas station, and every morning having a full tank of energy after charging up at night is incredibly convenient. The Model S is so smooth, quiet, and comfortable that driving it is a real pleasure, and both my wife and I find that after a long drive we feel more relaxed, less stressed, than we did in any of our previous cars, which were a diverse group: a Toyota Prius, Mini Cooper, Lexus LS400, Porsche Cayman and Subaru Forester. The acceleration of the Model S85 of course beats all those cars handily, including the Porsche, by the way.

The Model S user interface, which is primarily the 17-inch touchscreen, is very well designed and easy to learn and use. We find the seats to be quite comfortable, and the massive storage space in the trunk and the 'frunk' makes it easy to take whatever we want to with us on long trips.

The internet connectivity included with the car means that software updates just show up as they are rolled out (no more trips to the dealership to get an upgrade), and the on-board navigation maps are automatically updated, plus the Google Map navigation on the 17-inch screen is always up-to-date. Since I've had the car there have been new features added through the software, including a Hill Assist feature that holds the car in place for a moment after the brakes are released when on an incline.

To me the only negative of the car is that it is so big (which of course gives it all that storage space we like!). I've gotten used to the size, but strongly recommend the Tech Package option that includes the Parking Sensors, as they are very helpful when parking a car of this size.

The size is the only real negative for me. The list of positives is lengthy: performance, handling, comfort, roomy interior, quiet cabin, stunning looks inside and out: all round it is the finest car I have ever owned. The electric drivetrain means far less maintenance and the 8-year/infinite miles drivetrain warranty is something that no other car manufacturer can match.

I had an issue with the 80 amp High Power Wall Connector (an option: most owners charge at 40 amps / 240 volts at home), where it stopped working. Someone from the local Tesla service team came to my house that day and replaced the cable; no problems since. Several months ago I got an email from Tesla saying they wanted to proactively replace the driver display screen (in a typical car, that would be called the gauge cluster) because there had been problems with some of them, though I had not had a problem. I took the car in and waited an hour while that was done. The chrome trim pieces at the rear of each passenger window are a little loose and I will get those fixed when I get around to scheduling a date to have the Triple Underbody Shield installed (which was added to cars after two separate collision accidents in late 2013 resulted in a battery fire. The car alerted the driver to stop and exit the vehicle before the fire started; there have been no fires since). However, I am in no rush to do that because I think the battery fire issue was overblown: every day hundreds of ICE (Internal Combustion Engine) cars catch on fire and no one pays any attention.

I highly recommend the Model S to everyone I talk to about the car. The base 60 kW version with no options is a very nice car but I think the Tech Package option is really worth it (see the Tesla website for the latest information on what is included in that option, as the feature list keeps expanding) and the 85 kW battery is nice to have if you can afford the extra cost. I did not get the Air Suspension or the Panoramic Roof (sunroof) options and don't miss them. The Air Suspension would be useful if you have a steep driveway.

With the growing network of Supercharges, the Model S can serve as a primary or sole vehicle. No other EV can do that because all other EVs are basically commuter or city cars only. My Model S ownership experience has been such a positive one that 10 months later I bought a 2008 Tesla Roadster. My household is now 100% EV.

Martin Ingles – Mitsubishi i-MiEV — United Kingdom

I have a great deal of interest in the environment and ways of reducing our greenhouse gases. I work in a related industry, designing wind-turbine generators, and I have even taken steps with installing my own solar hot water system. So the thought of owning an electric car for a year seemed the natural next step.

I must admit that I did not expect the car to be as good as it is. Rather than something a little limited in terms of speed and range, it is actually quite nippy and has had quite enough range for the duties of a second car.

When we first had the i-MiEV we were quite anxious about range. We were very conscious that the car could only do 80 miles per charge, but had not really considered how far we travelled every day.

Having driven a conventional car with a range of 300–400 miles on a tank, I have never had to think or plan what journeys I would be doing in a day, and so it took some time to get used to thinking about when to charge, or who (between my wife and me) would have the i-MiEV and who would have the other car.

What was also quite disconcerting was that as soon as you turn on the heater in the car, the range drops by around 10 miles (16 km). This, along with concerns about range, prompted us to have the car on charge almost all of the time it was parked on the drive.

On a couple of occasions I did push the range a little. When we had just got the car, and the batteries had not been bedded in, I ran the charge down to one or two miles to try to stretch the battery and speed up the bedding-in process. This seemed to work quite well, as the range did increase quite rapidly after subsequent charges.

On one occasion I miscalculated how far I had to travel and ended up driving the last three miles with the range meter on zero and flashing at me. In the last half mile a tortoise symbol came up on the dash and the power to the drive reduced quite significantly and so I crawled home – but I got there! As we have got used to the car we have become more relaxed about keeping the battery charged and when we are only making local trips we often only charge every other day.

Driving the car is fun. Everyone who has had a ride in it has commented on how quiet it is, and yet how good the acceleration is – especially from a standing start.

My verdict on the car is good. I think it is great, and despite the current energy mix going into the national grid, I think that it is good for the environment. All things considered, I am pleased and proud to have been an early user of a Mitsubishi i-MiEV.

Michael Walsh – Nissan LEAF　　　　　California, USA

I have been watching the trials and tribulations in the electric vehicle world for approximately fifteen years now, ever since becoming aware of the General Motors EV1 in the mid to late 1990s. Back then, the EV1 leased for a fair amount of money. Certainly too much for a young man who was yet to realize his full earning potential. Not to mention one with a job requiring an unpredictable amount of driving on any given day. While I thought it was a very cool idea, it was not very practical for my set of circumstances at the time.

However, I kept up with goings-on in the marketplace, watching each successive attempt at a commercially viable electric car arrive on the scene, and became ever more determined to make the first affordable, highway-capable one, mass-marketed by a major manufacturer, a part of my life. This is where Nissan and the LEAF come into the picture, winning my allegiance by being the first.

I have to be honest that I was never very big on the Nissan's styling – it is a bit on the weird side by American standards. However, seeing all manner of strange vehicle shapes on a vacation to Europe convinced me that I could live with the slightly oddball look, especially since it met the ultimate goal of decreased dependence on foreign oil for my family.

Nevertheless, it was not that cut and dry. The vehicle still needed to meet my daily needs, preferably without having to mess around with finding places to fuel while out and about. I wanted assurance that it was perfectly capable of driving at least 60 miles at full highway speeds. Fortunately that turned out to be the case, and I can comfortably get to the office and back with at least 25 miles to spare. Not that I left it to chance, ingratiating myself as I did with the Nissan public relations people to gain preferred access to their press demonstrator vehicles. I already had a pretty good idea of what the vehicle could do, well in advance of receiving my own.

The night before delivery of my LEAF three weeks ago, my wife and I celebrated by opening a bottle of inexpensive champagne and watching Chris Payne's *Who Killed the Electric Car*. It seemed a fitting thing to do. The next day, bright and early, I took my place as a foot soldier in the new EV revolution.

So what has the LEAF been like to live with these last few weeks?

The car itself is remarkable. It is well thought out and the engineers and designers have not left much of anything to chance. It drives well: acceleration to highway speeds is effortless and handling (with the exception of the low-speed steering that many Europeans will find a little light) is very good.

There are some quibbles I think many long-time electric vehicle enthusiasts will have. The absence of an accurate State of Charge meter and more aggressive regenerative braking are just two that come immediately to mind. But for the most part, there are going to be few complaints by those who understand what a major step forward this car represents at this particular point in history.

Angela Boxwell – Mitsubishi i-MiEV   United Kingdom

As a mum, I like to know that my children are comfortable, safe and happy. We have been using the i-MiEV as our everyday car for the past year. We live in a village and our nearest town is six miles away and the children's school is a five-mile round trip.

We therefore need a car suitable for lots of short journeys. Most days I find myself zipping between school, shopping and taking the children out. The i-MiEV fits in with our family life perfectly.

On average our daily mileage is 22 miles. This includes driving in villages, along main roads and dual carriageways. The i-MiEV can cope with any situation and has the acceleration and speed to keep up with other traffic. The children love how quiet the car is, especially when sat in traffic. I like the feeling of turning up to school in my environmentally friendly car when a lot of the parents are turning up in their big four-wheel drives.

It is more relaxing to drive due to the lack of vibration and noise from the engine. Most of the time I find myself using eco driving techniques, changing between *Comfort* mode for driving and *Boost* mode for getting the best out of the regenerative braking. If I want to pull away quickly I use *Boost* mode, knowing that the car is very responsive.

The children have plenty of room in the back and it can also seat two adults in the back quite comfortably. My father-in-law is 6 feet 5 inches tall and he can comfortably sit in the back of the car.

I use our i-MiEV for the weekly shop and there is plenty of luggage space. I have even taken three children and their luggage for a couple of nights away.

I normally charge our i-MiEV up every two or three days unless I know I need a full charge for the following day. For us, charging at home is more convenient than having to find a fuel station: living in the countryside means I used to drive five miles out of my way just to get fuel.

I would recommend the i-MiEV to other families as their everyday car. It has fitted in perfectly with our requirements.

Damian Powell – Nissan LEAF — United Kingdom

My old car was becoming expensive. Every service required unexpected work on top of the work that I knew about, the fuel economy was getting poorer, the mod-cons were, well, not modern, and it didn't have any cup holders. So I decided it was time to get a new car, and by new I mean brand new, for the first time ever. I've always been interested in the idea of green motoring and after going for a ride in a friend's LEAF I decided that now was the time to try.

My initial impression of the LEAF was that it is a very sharp car to drive. That is, it feels very tight, very responsive while at the same time being relaxed and comfortable. I can honestly say that I enjoy driving this car. Unfortunately for me, my partner drives the car more than I do because I work from home, and she has a relatively long commute; it seemed silly for me to have the more economical car at home while she spends money on a relatively inefficient petrol driven car. The upside is that I get to drive it at the weekend and I look forward to doing so.

What would I say are the major pros of owning an EV? My trips to service stations are now far less frequent. The only time I go there is to check wiper fluid, or to get fuel for me. The car itself is very comfortable, well specified and intelligently designed. I also think that the car has helped me to become a better driver. This is because of all the tools that the LEAF has to guide economical driving, which make it easy to understand that bad driving is often inefficient driving. Since driving the LEAF, I find that I drive my ICE car more considerately.

There is one major drawback with an EV. It's the obvious one: range. Now I'm not saying that range is poor on the LEAF. In fact, the range of the LEAF is quite good and I very rarely need to use charging-points away from my home. However, the range can be severely impacted by adverse conditions. Wind, rain, cool temperatures, high speeds, and steep inclines – these can all be detrimental. The real problem is not that the range is affected, but that it is very difficult to predict how badly it will be affected.

The other major con is the availability of charging-points. It's fair to say that the availability of charging-points has probably doubled in the year since I have had my LEAF. That's certainly true when talking about rapid chargers. Other drivers are inconsiderate of EVs, though, and often park in spots reserved for EV use.

I love the LEAF. I love driving it. I love the fact that I'm not having a detrimental effect upon the environment when I drive it (I get my electricity from a green energy supplier). However, the infrastructure is still not there to make it a carefree driving experience (although it has nearly doubled in just the last year) and legislation is not yet there to treat rare, EV charging spaces with the same reverence as disabled parking spaces. I think these things will continue to change over the next couple of years though, and as battery technology improves and ranges increase, I expect that EV driving will become normal. That is to say, we won't have to pay special attention to the weather, we won't have to drive slower than all of the other vehicles, and we won't have to worry about being caught short because there'll be rapid chargers at all the major service stations. My next car will also be an EV.

Electric Cars and the Environment

Few subjects are more emotive than the environment. On the subject of the environmental benefits of electric cars, everybody seems to have an opinion. Claims and counter claims are made and after a while it becomes difficult to distinguish folklore from fact.

I am not going to use this book to jump into the climate change debate. Instead, I will do my best to explain the environmental impacts of electric cars and how they compare with conventional combustion engine cars, and leave the debate on global warming to others.

Before we begin

I have gone into this section in a fair amount of detail. This is the only way to explain the *big picture* about the environmental impact of road transport, both electric and otherwise.

Some people may not want this level of information. I have tried to keep it interesting and not get embroiled in the mathematics, but feel free to skip to the end of the chapter and just read the conclusions if you want.

Others need to understand the environmental case for and against electric cars in great detail and will not be satisfied until they have got it. I hope I have gone into enough detail to satisfy everyone. I have included references to all the studies and reports I refer to, so you can track the information back to its source.

Comparing electric cars with conventional cars

Electric car enthusiasts are always keen to point out the fact that their cars do not emit any pollution where they are being used. Detractors point to the coal-fired power station generating the electricity in the first place.

Both groups are making a valid point but, taken in isolation, both groups are wrong. Without looking at the whole picture, no fair assessment of the relative merits and disadvantages of different technologies and vehicle types can be made.

Why the current comparisons fail

The European Commission has defined a standard way for measuring the emissions from cars, based on their carbon dioxide emissions from the exhaust of the car. This is measured in grams of carbon dioxide per km travelled (CO_2 g/km).

The US Environmental Protection Agency has a similar measurement, with the results shown in grams of carbon dioxide per mile travelled (CO_2 g/mile).

Both of these measurements are 'tank-to-wheel' measurements: a measurement of emissions from the point the fuel has been pumped into the fuel tank to the point where the energy is used.

Of course, electric cars benefit significantly from this measurement because by themselves they do not pollute at all: all the pollution happens at the power station where the electricity is generated.

However, in the same way that a 'tank-to-wheel' measurement does not measure the true carbon footprint of using an electric car, neither does it measure the true carbon footprint of using a conventional combustion engine car.

For a conventional car, the carbon footprint for extracting, refining and transporting the oil needs to be taken into account, not just the carbon footprint for the emissions coming out of the tailpipe.

For an electric car, the carbon footprint for getting the raw fuel, transporting it to the power station, generating the electricity and 'delivering' it to the plug needs to be taken into account.

These measurements are called 'well-to-wheel' measurements. In order to be able to make a true comparison between electric cars and combustion engine cars, we need to be able to identify this well-to-wheel calculation for both oil use and electricity use.

How to create a proper comparison

There are several measurements that need to be considered when comparing the environmental impact of a conventional car with an electric car:

- Air pollution – at the tailpipe *and* at the electricity power station.

- How the energy is produced and transported – all the way back to where the oil is pumped out of the well and coal is extracted from the mine.

- Fuel economy.

- The environmental impact of batteries.

- Vehicle manufacturing and distribution.

- Recycling the vehicle at the end of its life.

I have structured this chapter to discuss these measurements separately. At the end of the chapter I then pull the threads together to provide a suitable comparison between combustion engine cars and electric cars.

An acknowledgement to motor manufacturers

There is a tendency for environmental groups to cast the big motor manufacturers, along with 'Big Oil', as the environmental demons of the known world.

You do not need to look far on the internet to start finding conspiracies about motor manufacturers and big oil producers working in collusion; or to see pictures of big, old factories pumping out high levels of emissions while all the time building bigger and more powerful SUVs.

Cars are bigger and heavier than twenty years ago. This has been driven by customer demand for larger cars with more power and higher levels of comfort and safety. Despite this, fuel economy, fuel emissions and engine efficiency have consistently improved during this time.

In fact, car manufacturers have done more to improve the environmental performance of their products over the past thirty years than any other industry.

More recycled materials are being used to build new cars, making a significant reduction in raw material usage. In addition, around 95% of a new car can now be recycled at the end of its useful life.

Some of this improvement has been down to legislation. However, manufacturers have also been taking the initiative in making a better quality product that is more environmentally friendly than ever before. The vast majority of car manufacturers consistently deliver environmental improvements in their products years ahead of legislation taking effect.

Motor manufacturers are also significantly improving the efficiency of internal combustion engines, producing vehicles that are far more fuel efficient than ever before. Even in the past four years, since the first edition of this book was written, there have been some significant improvements in fuel efficiency figures.

There is still a long way to go. By no means am I saying that motor manufacturers are perfect. Yet it is still a point worth making: above almost every other industry, car makers have been leading the way to improve the environmental efficiency of their products. The evidence suggests they will continue to do so for many years to come.

Air pollution

Whatever your opinions on climate change, there is no doubt that we suffer from over-pollution in our towns and cities.

In Germany, it is estimated that over 65,000 people die prematurely every year as a result of excessive air pollution. Public Health England estimated that 28,000 people died prematurely in the United Kingdom due to air pollution during 2010 and that air pollution is responsible for 5.3% of all deaths in the over-25s[13]. Across Europe, air pollution reduces life expectancy by an average of around nine months. In some European countries and in some highly congested cities, the average is closer to 1–2 years[14].

Worldwide, two million people die each year as a result of excessive air pollution[15]. Tens of millions of people suffer from pollution-related illnesses, such as heart and lung diseases, chest pains and breathing difficulties[16] [17]. Air pollution is now seen as a major public health issue and not just an environmental concern.

Air pollution from road traffic

Transportation gets a lot of criticism for creating air pollution. There is no doubt that it is one of the major contributing causes of air pollution, but it is by no means the only one. Industry and homes all create air pollution, as of course do power stations that generate electricity, so why does traffic pollution get all the attention?

There are many reasons why traffic pollution has been singled out by scientists and policy makers as the most significant pollution issue:

- Worldwide, industrial and domestic pollutant sources are generally improving over time, whilst pollution from traffic is becoming worse[18].

- Tiny particles within vehicle exhaust, known as particulates or particulate matter (PM), are particularly dangerous when breathed in. Particulate matter penetrates

---

[13]Source: Estimates of Mortality in Local Authority Areas Associated with Air Pollution. Published by Public Health England, April 10th 2014.

[14]EC study – Thematic Strategy on Air Pollution, COM(2005) 446 final, 21.09.2005.

[15]World Health Org: Fact Sheet No. 313 – Air Quality and Health.

[16]Air Pollution-Related Illness: Effects of Particles: Andre Nel, Department of Medicine, University of California. Published by the AAAS.

[17]The Merck Manual for Healthcare Professionals: Pulmonary Disorders.

[18]EUROPA research into air quality at the European Commission MEMO/07/108, 20/03/2007.

deep into the lungs and in some cases directly into the bloodstream; it has the potential to affect internal organs.

- Vehicle exhaust contains both volatile organic compounds (VOCs) and nitrogen oxides (NO_x). In the presence of sunlight, this creates a complex chemical reaction, producing ozone. When inhaled in small amounts by relatively healthy people, ozone can cause chest pain, coughing, shortness of breath and throat irritation.

- Carbon monoxide emission from traffic exhaust is poisonous. It enters the bloodstream and reduces oxygen delivery to the body's organs and tissues.

- Toxic Organic Micro Pollutants (TOMPs), produced by the incomplete combustion of fuels, comprise a complex range of chemicals that can be highly toxic or carcinogenic. TOMPs can cause a wide range of effects, from cancer to reduced immunity of the nervous system, and can interfere with child development[19]. There is no 'threshold dose'. Even the tiniest amount can cause damage to internal organs[20].

- Traffic pollution is the number one cause of particulate matter and carbon monoxide emissions in the world and a major contributor to VOC, nitrogen oxide and TOMP emissions.[21]

- Heavy traffic can create very high levels of pollution in areas where there is high population density, such as in major towns and cities. This pollution results in very high human exposure to these emissions.

---

[19]Information sourced from the Toxic Organic Micro Pollutants (TOMPS) Network web page at http://uk-air.defra.gov.uk/data/tomps-data.

[20]Air Pollution in the UK: 2007. AEA Energy and the Environment, commissioned by DEFRA and the Devolved Administrations, UK.

[21]University of Strathclyde Energy Systems Research Unit: Environmental Pollution from Road Transport.

Diesel pollution

Emissions from diesel vehicles are particularly nasty. Carbon dioxide emissions are comparatively low, but nitrous and other particulates in the emissions are harmful to human health and are a significant issue with local air quality in many cities.

Diesel exhaust (known as diesel particulate matter or DPM) causes short-term symptoms such as dizziness, headaches, nausea and breathing difficulties. Exposure can lead to chronic health problems such as cardiovascular disease and lung cancer.[22]

In 1998, the California Air Resources Board identified diesel particulate matter as a 'toxic air contaminate', based on its potential to cause cancer, premature death and other health problems. The American Lung Association estimates that DPM causes 4,700 premature deaths annually in nine of America's major cities. The discussions of the UK's Governmental Committee on the Medical Effects of Air Pollution (COMEAP) suggest that DPM could be significantly worse than had previously been believed[23].

Diesel particulate matter is particularly dangerous as many of the particles are very small, making them almost impossible to filter out and very easy for human lungs to absorb.

Four things have been introduced to reduce DPM pollution:

- Fuel companies have introduced ultra-low sulphur diesel (ULSD). Otherwise known as 'clean' diesel, ultra-low sulphur diesel has approximately 3% of the sulphur found in normal diesel. This reduces the amount of soot found in diesel emissions, with only a very slight impact on peak power and fuel economy.

---

[22]University of Strathclyde Energy Systems Research Unit: Environmental Pollution from Road Transport.

California Environmental Protection Agency: Air Resources Board. Diesel Health Effects.

[23]COMEAP Minutes: 5th November 2013 meeting.

- Car manufacturers have developed their modern diesel engines to become significantly more efficient at burning their own emissions, thereby reducing the quantities of DPM being released into the atmosphere and significantly decreasing carbon monoxide emissions.

- Modern diesel engines have complex filtration to filter out the larger particles.

- Some of the very latest diesel engines now have a urea-injection system to inject urea into the exhaust in order to convert harmful nitrogen oxide gas into ammonia and nitrogen.

    Ammonia is, however, a toxic gas that can be fatal if inhaled directly and is a highly reactive environmental pollutant. Yet it is still better than the nitrogen oxide it replaces.

These advances are very welcome and will undoubtedly save thousands of lives around the world. Yet despite the advances, diesel pollution remains a significant concern.

Compared to an equivalent petrol (gasoline) engine, a modern 'clean' diesel engine is likely to produce 20–30 times more nitrous emissions, as well as particulate matter which the petrol/gasoline engine does not.

Unfortunately, the environmental ratings for cars are measured purely on carbon dioxide emissions and not the total pollutants. Diesel cars perform particularly well under this measurement. Yet if you take into account the hydrocarbon and the nitrogen oxides emissions from diesel engines, it is a very different story. A big, heavy Volvo V70 estate car with a 2½ litre petrol engine emits 201 mg/km of hydrocarbon emissions, while a tiny Fiat 500 city car with a 1.3 litre economical diesel engine with stop-start technology, puts out 484 gm/km[24]. Despite being half the size of the Volvo estate, the hydrocarbon emissions on the Fiat are over twice as high.

---

[24]Vehicle Certification Agency (VCA).

Unfortunately, many governments made a policy to promote diesel vehicle adoption in the 2000s, citing lower CO_2 emissions as a primary driver on doing so. Many governments now recognise this as being a mistake. Lord Drayson, a former science minister in the UK Government, has recently been on record as stating that the UK Government policy was wrong, saying that the health effects of the products of diesel are killing people in the United Kingdom[25].

What about biodiesel?

Biodiesel is often mixed with 'fossil fuel' diesel. In Europe, diesel purchased from a service station will typically have a 5% biodiesel mix. 15% and 30% mixes (called *Biodiesel B15* and *Biodiesel B30*) are now regularly available in many parts of Europe.

Only *Biodiesel B100* is 100% biodiesel. All other biodiesels are a blend of biodiesel and 'fossil fuel' diesel.

Biodiesel and some biodiesel blends are not compatible with all diesel engines. If you are planning to use biodiesel, check with your car manufacturer first.

Biodiesel can be made from many different sources. Traditionally, the most common source was plant oil produced from rapeseed, corn and maize. However, biodiesel production has been blamed for increasing food prices and causing starvation in the third world, and for increased deforestation in the Amazon rain forests.

Now, new second generation bio-fuels are used that are not made from food crops. Instead they use the waste from food crop production to make fuel. Other biodiesel is being made from algae, decaying plants, waste plastics and cardboard, and even landfill waste.

As conventional diesel becomes more expensive, biodiesel also has a role to play in ensuring a cost-efficient continuity of supply in the future. Worldwide demand for diesel is soaring, both for road transport and for shipping. Biodiesel mixes are going

---

[25] Interview on BBC Radio 4 Today programme, 1st October 2015, and reported on the BBC website http://www.bbc.co.uk/news/business-34407670

to become more important as a way to ensure long-term diesel costs are kept under control.

Biodiesel pollution

Theoretically, biodiesel is carbon neutral because the carbon dioxide emitted from burning the fuel is the same carbon dioxide that was absorbed by the plants as they grew. As plants die and decay, much of the carbon dioxide that they absorbed as they grew is released back into the atmosphere anyway, so converting the plants to biodiesel and then burning it is extremely carbon efficient.

Biodiesel has been touted by many as a clean fuel because of its significantly lower carbon dioxide and carbon monoxide emissions and its virtual eradication of sulphur from the exhaust. For this reason, there is no doubt that biodiesel can vastly improve air quality from traffic pollution.

However, nitrogen oxide emissions in biodiesel are much higher than in fossil diesel. Nitrogen oxide is almost 300 times more powerful as a greenhouse gas than CO_2.

This is one of the reasons why questions have been raised as to whether biodiesel has any benefit in terms of solving climate change issues. While the general consensus is that burning biodiesel rather than fossil diesel is advantageous in terms of environmental pollution, there is credible research to suggest that bio-fuels may actually increase greenhouse gas emissions rather than decrease them[26].

---

[26] Atmospheric Chemistry and Physics: N_2O release from agro-biofuel production negates global warming reduction. Atmos. Chem. Phys. Discuss., 7, 11191-11205, 2007.

Average petrol/gasoline and diesel emissions (per litre[27]) from fuel tank-to-wheel[28]:

	Carbon Dioxide	Carbon Monoxide	Nitrogen Oxide	Sulphur Dioxide
Petrol/ Gasoline	2,315g	140g	9.5g	Trace[29]
	5 pounds 1oz	5oz	3½oz	
ULS Diesel	2,630g	237g	37g	Trace[30]
	5 pounds 13oz	8½oz	1$^{1}/_{3}$oz	
B100 biodiesel	1,736g	117g	40g	Nil
	3 pounds 13oz	4oz	1½oz	

Please note: This table is based on average emissions for cars and light goods vehicles in the United States. The exact amount of carbon monoxide, sulphur dioxide and nitrogen oxide will vary significantly from one engine design to another and will also vary depending on how efficiently the engine is running.

---

[27]To convert these figures into emissions per gallon, multiply the figures by 3.79 for a US gallon; or by 4.55 for an imperial gallon.

[28]Source: US National Vehicle and Fuel Emissions Laboratory / BP Fuels.

[29]Sulphur-free petrol, containing less than 10 parts per million, became mandatory in 2009 across the EU. US figures may vary.

[30]Sulphur-free diesel, containing less than 10 parts per million, became mandatory in 2009 across the EU. US figures may vary.

A word about asthma and road traffic pollution

Over the past thirty years, there has been a huge rise in the number of asthma sufferers around the world. Many environmental groups claim that air pollution is the cause.

The figures on asthma are certainly shocking. In the United States, asthma rates have increased by 70% in the past thirty years and are up 160% in children under the age of five. 8.9% of all children in the USA are now diagnosed with asthma.[31]

In the UK, around 9% of all children and an additional 9% of all adults are currently receiving treatment for asthma[32], and similar trends can be seen across the whole of Europe and Canada.

Research into what is causing this large increase of asthma cases is still ongoing, especially in California, where new evidence has emerged that may yet link asthma with traffic pollution.

However, based on a number of studies carried out in Ireland, the UK, Canada, France and the USA, the general consensus in all these countries is that there is currently little evidence of a definitive link between the cause of asthma and air pollution.

There are many indicators that suggest that the causes lie elsewhere. Most asthma specialists believe that the cause is most probably in the home, where higher levels of insulation and heating over the past thirty years are creating the ideal environment for dust mites, with the asthma being caused by humans breathing in the waste products of these mites.[33] [34]. High exposure to household cleaning products is

---

[31] American Academy of Allergy, Asthma and Immunology.

[32] Asthma UK website.

[33] UK Department for Health Report: Asthma and outdoor air pollution, ISBN 011321958x.

[34] American Journal of Respiratory and Critical Care Medicine (1999; 159:125-29).

thought to have an impact too, and people with cleaning jobs have a higher than average risk of developing asthma[35].

Although air pollution may not be the cause of asthma, many people who already have asthma do report increased asthma attacks as a direct result of traffic pollution: this link between *asthma attacks* and air pollution is well known to specialists.

In a recent study in Wales, 65% of asthma sufferers said they suffered coughing fits and shortness of breath as a result of traffic congestion. Research in the USA has suggested similar figures[36].

When questioned, many asthma sufferers say they cannot walk or shop in congested areas because traffic pollution triggers breathing problems.[37]

Air pollution from electricity generation

The amount of emissions produced by electricity generation depends on a number of factors, including:

- What fuel is being used to generate the electricity (coal, coke, oil, gas, biomass)
- The quality of the fuel used (not all coal or oil is created alike)
- Where the fuel originates from (i.e. locally sourced or imported from elsewhere)
- The efficiency of the power station
- The filtration used to clean the emissions before release into the atmosphere

---

[35]Asthma UK research 'Cleaning Jobs linked to Asthma Risk', published 22 January 2013.

[36]American Academy of Allergy, Asthma and Immunology

[37]Asthma UK Opinion Research, 21st September 2007.

Around the world, air pollution from electricity power stations is reducing. In the United Kingdom, for example, emissions from electricity generation dropped by 11% between 2008 and 2009[38], and these emissions have fallen significantly again since then. Coal generation has dropped significantly as older power plants have closed and others have switched to burning biomass. Meanwhile, production from gas-fired power stations has dropped by almost 50% since 2010. Nuclear power is showing a slight increase whilst renewable energy, particularly from wind farms, is increasing by around 30% a year and now accounts for over 16% of all electricity generated in the United Kingdom[39].

Many other countries, including traditionally high-polluting nations such as Australia and India, are also reporting significant reductions in air pollution from electricity generation.

In the main, these improvements are not due to higher efficiencies from new power stations. The majority of these improvements are coming from existing power stations. These power stations are producing lower emissions than ever before as they switch to lower-carbon fuels, install particulate and sulphur filters and become more efficient at generating electricity.

Here are average figures for air pollution for each kWh of electricity generated by different types of power station:

---

[38]Digest of United Kingdom Energy Statistics (DUKES) 2010.

[39]Digest of United Kingdom Energy Statistics (DUKES) 2015.

Power station air pollution (per kWh of electricity generated)[40]

	Carbon Dioxide	Carbon Monoxide	Nitrogen Oxide	Sulphur Dioxide
Coal	990g 2 pounds 3oz	0.2g < 1/100th oz	2.8g 1/10th oz	2.7g 1/10th oz
Gas	400g 15oz	0.1g < 1/200th oz	0.4g 1/70th oz	Trace
Oil	640g 1 pound 11oz	0.4g 1/70th oz	3.2g 1/9th oz	1.5g 1/20th oz
Nuclear	16g ½ oz	Nil	Nil	Nil
Geo-thermal	122g 4½ oz	Nil	Nil	1.0g 1/28th oz
Hydro-electric	Nil	Nil	Nil	Nil
Wind turbine	Nil	Nil	Nil	Nil
Solar	Nil	Nil	Nil	Nil

Average air pollution for electricity generation

Every nation has a different mix of power stations to generate their power. The US, Australia and India all have a high dependency on coal. France uses nuclear power and Norway and Sweden generate most of their electricity using hydro-electric power

---

[40] I have various sources for this information – including discussions with representatives from EDF Energy, both in France and in the UK, E-ON and the UK National Grid, plus documented sources from the IEA Energy Technology Perspectives 2008 paper, the Laboratory for Energy and the Environment at the Massachusetts Institute of Technology and the Icelandic Energy Authority.

stations. As a result, there are significant differences in the average air pollution for electricity production from country to country.

The good news is that, around the world, carbon figures are coming down. I first created the table below in 2010, based on figures collated by the International Energy Agency. When I reviewed the figures at the end of 2014 for this edition of the book, every country, with the exception of Germany, Japan, Denmark and Pakistan, had managed to reduce their carbon footprint for energy production. In the main, this was because:

- Wholesale gas prices have increased by around 50%, making gas an expensive fuel to use to run power stations.
- Older coal-fired power stations are being decommissioned. Newer coal-fired power stations are more efficient, and many are being converted to run on biomass.
- We are now seeing a significant amount of electricity being generated by renewables, with a significant increase in energy production from wind power and solar.

Average power station air pollution by country[41]

Country	CO_2 emissions per kWh
Austria	215g (7½ ounces)
Australia	823g (1 pounds 13 ounces)
Canada	167g (6 ounces)
Denmark	315g (11 ounces)
France	61g (2¼ ounces)

---

[41]International Energy Agency Data Services/Carbon Trust.

Finland	191g (6½ ounces)
Germany	477g (1 pound 1 ounce)
Greece	720g (1 pound 9½ ounces)
Iceland	1g (1/28 ounce)
India	856g (1 pound, 14 ounces)
Ireland	427g (15 ounces)
Italy	402g (14 ounces)
Japan	497g (1 pound 1½ ounces)
Nepal	1g (1/28 ounce)
New Zealand	141g (5 ounces)
Norway	7g (¼ ounce)
Pakistan	409g (14½ ounces)
Spain	291g (10½ ounces)
Sweden	17g (2/3 ounce)
United Kingdom	430g (15 ounces)
United States of America	503g (1 pound 2 ounces)

The reason for Japan and Pakistan's increase in carbon footprint is down to natural disasters. The Fukushima Daiichi nuclear disaster, triggered by the tsunami in 2011, shut down one of Japan's largest nuclear power stations, whilst flooding in Pakistan has damaged two power stations, resulting in the use of lower efficiency emergency power generation.

Germany's higher carbon footprint is due to a policy change. As a direct result of the Fukushima incident, Germany moved away from nuclear power. This has resulted in a temporary increase in the carbon footprint of electricity generation, which it is claimed will be offset by increased investment in renewable energies over the next five years.

Full lifecycle emissions

The above table does not show the full story on carbon footprint or emissions. There are lots of 'hidden' emissions that need to be taken into account that the above figures do not include:

- Extraction and transportation of the fuel (such as coal and gas) to the power station
- Building the power station
- The day-to-day operation of the power station
- Transmission losses – i.e. electricity lost between the power station and the consumer
- Decommissioning the power station at the end of its life

We will look at these hidden emissions later on when we look at the *pit to plug* environmental impact for using electricity.

How our energy is produced

Whether we are using gasoline (petrol), diesel or electricity to fuel our cars, we need to understand how our energy is being produced in order to understand the full 'well-to-wheel' energy usage of our vehicles.

How gasoline, petrol and diesel are produced

Oil extraction

Crude oils can be found at varying depths in the earth's crust, either on land or under water. Oil is typically found trapped in sandstone, along with natural gas and salt water, often at high pressure.

The texture and chemical makeup of oil varies between oil fields. In some fields, the oil is like tar (so-called 'heavy crude'), whereas in others the oil is very thin and light ('light crude'). The chemical makeup varies, too, and is typically graded by its sulphur content. Low sulphur content oil is called 'sweet', while high sulphur content oil is 'sour'. Some crude oils are black or brown; others are red, green, yellow or orange.

The most valuable oil is 'light, sweet' crude oil as this yields a greater amount of automotive and aerospace fuels once refined.

Oil is released by drilling a well into the side of the formation. Tap the peak of the formation and you will only extract gas, tap too far to the side of the formation and you will only get salt water.

The oil normally pushes itself up out of the well because of the natural pressure found inside the oil seam. It often reaches the surface with quite some force.

As oil comes up out of the well, water is pumped back to replace the oil that has been extracted. This maintains the pressure levels within the well, easing the further extraction of oil.

Most oil fields have multiple wells, extracting both oil and natural gas. The oil and gas is either piped directly to a refinery or loaded onto ships for onward transportation to a refinery.

Along with oil and gas, sand, salt and water are extracted from the well. As the oil fields get older, more and more sand, salt and water are extracted. In some oil fields, 99% of the extracted liquid is salt water. The oil is separated from the salt water and sand as part of the extraction process. The more filtering that needs to be done, the higher the cost of extraction, to the point where it becomes uneconomic to continue operating near-exhausted oil fields.

Getting fuel from crude oil

Crude oil cannot simply be pumped into a car engine and used as fuel. Crude oil is a mixture of different components (called fractions). Some fractions are highly combustible, light and runny; some fractions are thick, heavy, inflammable tar-like sludge and other fractions are somewhere between the two.

As crude oil, these different fractions are all mixed together into one liquid. To make crude oil into a useful material, it first needs to be separated, converted and treated.

This is the work that is carried out at an oil refinery.

An oil refinery transforms crude oil into lots of different products:

- Bitumen (tar) is a heavy, inflammable liquid, used for the production of road surfaces and roofing materials.

- Fuel oil is another heavy liquid, used in oil-fired power stations.

- Lubricating oils are used for making oil-based polishes and grease.

- Diesel.

- Kerosene is used in the production of jet fuel, paraffin and heating oil.

- Petrol / gasoline.

- Naphtha is used in the production of plastics.

- Petroleum gas is used for making liquid petroleum gas (LPG) and bottled gases.

Hot crude oil is pumped into a distillation tower, where it continues to be heated. As the different densities of oil heat up, the crude separates into fractions according to their density and boiling point, which allows them to be extracted.

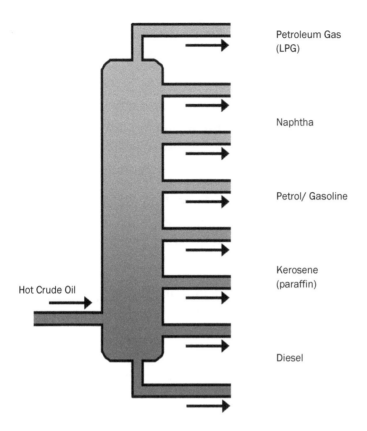

Figure 1 - A diagram of a distillation tower, where crude oil is separated into different categories of oil by heating it.

Lighter oils, such as LPG and petroleum, evaporate under heat and rise to the top of the tower. Kerosene and diesel remain near the middle and heavier liquids, such as lubricating oil and tar, remain at the bottom. These different fractions are then tapped off and taken off for further treatment.

Only some of the crude oil is suitable for road fuels, with the rest being used for other purposes. However, the demand for petrol and gasoline is so high that oil refineries need to be able to create more road fuel from crude oil than is possible through distillation alone. To do this, some of the low-value heavier oils go through an energy-intensive process known as catalytic cracking. Oils are pumped into a reactor and heated to an intense heat and pressure. Using a catalyst, catalytic cracking then converts most of the heavy fractions into lighter fractions such as petroleum and diesel.

Some oil refineries use another energy-intensive process called coking. This is used to break down very heavy fractions (tar and lubricating oil) into lighter oils, leaving a residue of coke – a hard, high-carbon, coal-like substance that is used as an industrial fuel and can be burnt like coal at many coal-fired power stations.

Once petrol, gasoline and diesel have been extracted from the crude oil, they undergo further treatments. These treatments include blending, purifying and adding additives to improve performance. Petrol companies often blend their fuel to ensure the best performance in different weather conditions.

On average, oil refineries are around 88% efficient at converting crude oil to refined oil[42]. This is an extremely high level of efficiency for any manufactured product.

'Well-to-service-station' efficiency for car fuels

According to the United States Department for Energy, the average 'oil well to service station' efficiency for car fuels is 83%[43].

This figure includes the transportation of crude oil to the refinery, the refining of the oil and the transportation of the fuel to the service station.

In research for this book, I have come across a number of credible claims that refining oil is significantly more energy-intensive than this and that the carbon footprint of fuel is therefore much higher.

While I have not been able to get any carbon figures directly from an oil refinery, the energy consumption and production figures of the oil refinery sector is a matter of

---

[42]Source: Estimation of Energy Efficiencies of U.S. Petroleum Refineries, Michael Wang. Center for Transportation Research. Argonne National Laboratory. March 2008.

[43]Source: DOE. Electric and Hybrid Vehicle Research, Development and Demonstration Program; Petroleum-Equivalent Fuel Economy Calculation.

public record in the United Kingdom. Although it is not possible to give accurate assessments of efficiencies within a refinery based on the very basic level of information available, there is sufficient information to suggest that the figures provided by the US Department for Energy are reasonably accurate.

In the same way, finding specific information for the carbon footprint for refining diesel has proved difficult. Part of this is down to the varying makeup of crude oils depending on its source: heavy, sweet crude can be broken down into diesel more efficiently than petroleum. Light, sweet crude can be broken down into petroleum more efficiently than diesel.

However, while certifiable data about diesel has proved elusive, I have been told by a number of industry experts that, on balance, the emissions caused by the refining of petroleum and diesel are comparable. For this reason, I am using the same figures for both.

Based on this calculation, you can divide the average fuel emissions by 0.831 – or multiply by 1.205 – in order to calculate the approximate 'well-to-wheel' emissions of petrol, gasoline or diesel-powered cars:

Average petrol/gasoline and diesel CO_2 emissions (per litre[44]) from the oil well-to-wheel[45]

Fuel	CO_2 emissions per litre
Petrol/Gasoline	2,789g (6 pounds 3½ ounces)
ULS Diesel	3,168g (7 pounds 1 ounce)

[44]To convert these figures into emissions per gallon, multiply the figures by 3.79 for a US gallon; or by 4.55 for an imperial gallon.

[45]Source: Discussions with representatives from BP Global and figures from the US National Vehicle and Fuel Emissions Laboratory.

Where our electricity comes from

As already discussed on page 137, our electricity comes from various sources and each source has a very different carbon footprint.

Managing these sources and ensuring that electricity supply always matches with demand is not a straightforward process. If supply does not match demand, the result is either a brownout or a blackout. A brownout is a voltage loss that causes lights to flicker or dim and electronic equipment to reset. A blackout is a complete power failure.

Electricity is delivered from power stations to homes and businesses through power grids. These grids can either cover an entire country, such as in the United Kingdom or France, or can be regional, as in the United States. In both cases, grids can have interconnections between them, allowing regions or countries to export and import power to cope with peaks and troughs in different regions at different times.

Some power stations run constantly, while others are adjusted to cope with demand. Nuclear power stations, for example, run at a constant output and as a consequence are often used as a 'base load'. In the United States, coal-fired power stations work in a similar way, although in Europe, newer coal-fired power stations have been improved to allow for more flexible power generation.

Gas and oil-fired power stations can adjust their output to provide greater or lesser power depending on demand. In times of low demand, these power stations may spend some of each day generating no power at all.

This means that it is not always obvious where your electricity is generated from. You could live next to a coal-fired power station but outside of peak times your actual electricity may actually be generated by a wind turbine hundreds of miles away.

Your power may not even come from your own country. For example, Canada and Mexico supplies parts of the United States with electricity, and the whole of Europe now has electricity interconnections between different countries.

Theoretically, the longest cost-effective distance for transmitting electricity is 7,000 km (4,300 miles)[46]. At present, all transmission lines are considerably shorter, but the likely emergence of renewable power super grids spanning entire continents in the next few years means it is not inconceivable that one day a person living in Scotland could be using solar electricity produced in the Sahara.

Demand for electricity goes up and down at different times of the day. In the dead of night, there is very little demand for electricity, while during the daytime and evening, demand is quite a bit higher.

A typical day in the UK for electricity demand would look something like this:

- Demand for electricity is very low throughout the night and early morning.

- There is a jump in power demand at around 8 o'clock in the morning.

- The demand then rises gradually throughout the morning, often with a mini-peak at around 10:30 am.

- Demand gradually rises during the afternoon, jumping up considerably at around 4 o'clock.

- Peak demand starts at around 5 o'clock in the evening and continues until around 7:30 pm, after which the demand for power gradually decreases, dropping rapidly again after 10 o'clock at night.

In the UK, the majority of the 'base load' (i.e. the minimum amount of power required on the national grid) is provided by gas, coal and nuclear power stations.

The UK 'peak load' electricity, which handles the different levels of demand throughout the day, is provided by gas, coal, hydro-electric, oil and pumped-storage power stations (a description of each of these is included later in the chapter).

---

[46]Global Energy Network Institute library paper: Present Limits of Very Long Distance Transmission Systems.

Wind turbines contribute power depending on the amount of energy they are producing at any one time. As we cannot rely on the wind blowing at the right time, it is difficult to factor wind energy into managing peaks and troughs in energy demand.

Coal-fired power stations

Coal-fired power stations produce the highest emissions of all for generating electricity. Worldwide, approximately 41% of all electricity is generated from coal-fired power stations[47], while around 49% of electricity in the United States is generated from coal.

Coal has achieved this position of dominance because of price and flexibility. Coal-fired power stations can use cheap coal to generate electricity at a price that is difficult for other fuels to compete with. At the same time, coal-fired power stations are reasonably flexible with their power output. When demand increases, they can boost their power generation accordingly and drop back again when demand decreases.

According to the World Coal Institute, if we continue to use coal at our current rate, we will have enough to last us another 130 years.

Like oil, the makeup of coal varies depending on where it is mined from. Coal is an organic rock made up of varying combinations of carbon, hydrogen, oxygen, nitrogen and sulphur.

Coal is 'ranked' based on its chemical and physical properties. Low-ranked coals are low in carbon but high in hydrogen and oxygen, while high-ranked coals are high in carbon but low in hydrogen and oxygen.

Low-ranked coals are less efficient to burn than high-ranked coals, creating more ash and soot. As a consequence, they cost less to buy. For this reason, most coal-fired power stations are designed to use the lowest ranking of coal available.

---

[47]Source: World Coal Institute.

Despite the huge environmental impact of mining coal, the actual carbon impact for extracting and transporting the coal required to generate 1 kWh of electricity is actually very low:

- Mining adds around 7 g/kWh of carbon to the footprint of the coal from an open cast mine, or around 14 g/kWh of carbon from a deep underground mine.
- Transporting coal by rail adds around 7 g/kWh.
- Shipping coal from one continent to another adds a further 2½ g/kWh.

Compared to the environmental impact of actually burning coal, the impact of mining and transporting coal is very low. Friends of the Earth say that all these processes equate to an additional 17 g/kWh of CO_2 for electricity generated by coal in the United Kingdom.

According to the Union of Concerned Scientists, in an average year, a typical coal-fired power station emits:

- 3,700,000 ton of carbon dioxide (CO_2)
- 10,000 ton of sulphur dioxide (SO_2)
- 500 ton of particulate matter (PM)
- 10,200 ton of nitrogen oxide (NO_x)
- 720 ton of carbon monoxide (CO)
- 220 ton of volatile organic compounds (VOCs)
- Small amounts of mercury, arsenic, lead, cadmium and uranium

In addition to the air pollution, the average coal-fired power station creates around 125,000 tons of ash and 193,000 tons of 'sludge'. This waste includes toxic substances such as arsenic, mercury, chromium and cadmium. In the United States, 75% of this waste is disposed of in unmonitored on-site landfills[48].

In the United States, the CO_2 emissions from coal-fired power stations average out at 990g CO_2 /kWh (2.18 pounds CO_2 per kWh). US strategy to reduce the carbon footprint of coal-fired power stations is focused around carbon capture technologies, where the carbon is stored underground. This is an expensive technology that is in its infancy. Whilst there is the potential for significant carbon savings using carbon capture, much of this will take many years to come to fruition[49]. In the meantime, very few of the improvements that we are seeing in European and Asian coal-fired power stations are being replicated in North America.

Coal is unlikely ever to be a truly clean fuel. It will remain a core fuel for electricity production in many countries for many years to come. In the meantime, any attempt to clean the emissions from coal-fired power stations is to be encouraged.

Burning biomass at coal-fired power stations

There is a lot that can be done to improve coal-fired power stations. One of the best examples of this is Drax Power Station, a coal-fired power station in the United Kingdom that went into operation in 1973.

In 2009, the largest single source of pollution in Europe was the Drax Power Station in North Yorkshire, United Kingdom. Its generating capacity of almost 4,000 megawatts of electricity was, and is, the second highest of any power station in Europe, producing around 7% of the United Kingdom's electricity.

Thanks to its pioneering work on biomass, Drax is now the most carbon efficient coal-fired power plant in the United Kingdom.

---

[48]Union of Concerned Scientists: Environmental impacts of coal power. June 30th, 2014.

[49]Coal-Fired Power Plants in the United States: Examination of the Costs of Retrofitting with CO_2 Capture Technology, Revision 3. January 4th, 2011. DOE/NETL 402/102309.

In 2009, Drax Power Station emitted 20,482,713 tons of carbon dioxide, 26,900 tons of sulphur dioxide, 38,190 tons of nitrogen oxide and 450 tons of particulate matter. 1,336,000 tons of ash was created by the plant, of which 780,000 tons were recycled and sold to the construction industry and 556,000 tons was placed in landfill[50]. In addition, an estimated 249,000 tons of carbon dioxide emissions were generated as a result of transporting coal to the power station.

In 2009, Drax was five years into a project to burn biomass at the power station instead of coal. Biomass – effectively waste product from the forestry and agricultural industries – releases some of the carbon from plant material that was absorbed by the plant while it was growing. Whilst it would be unfair to claim that this makes biomass entirely carbon neutral, it is significantly lower carbon than coal. Drax estimated that in 2009 the use of biomass reduced emissions by 581,896 tonnes.

Burning biomass is just one area where Drax has been attempting to improve pollution levels. The plant has also had considerable success in reducing sulphur emissions over the past five years and has recently won a €300 million grant to fund a carbon capture scheme that will bury carbon emissions from its coal-fired units into a depleted gas field in the North Sea to the east of the United Kingdom.

Since 2009, Drax has increased biomass burning considerably. In September 2012, the plant announced that it was fully converting three of its six generating units to biomass by 2017. At time of writing, the first two generation units have been converted and is functioning correctly. The third will go live later this year[51]. In the last year alone, biomass generation has increased from 7.9TWh to 11.5TWh and is now responsible for over 40% of its power generation.

According to Drax itself, the average CO_2 emissions for electricity generated at the power station have dropped considerably over the past few years as a result of its move from coal to coal/biomass:

---

[50]Drax Power Ltd Environmental Performance Review 2008.

[51]Drax Power Ltd Annual and Interim Reports – 2009–2014.

Year	Grams of CO_2 per kWh of electricity produced
2006	840g CO_2/kWh (1 pound 13½ ounces)
2008	818g CO_2/kWh (1 pound 13 ounces)
2009	815g CO_2/kWh (1 pound 13 ounces)
2010	784g CO_2/kWh (1 pound 11½ ounces)
2011	760g CO_2/kWh (1 pound 10½ ounces)
2012	785g CO_2/kWh (1 pound 11½ ounces)
2013	725g CO_2/kWh (1 pound 9½ ounces)
2014	629g CO_2/kWh (1 pound 6 ounces)
2015	595g CO_2/kWh (1 pound 5 ounces)

These figures look much more impressive when shown as a graph, clearly showing that switching from coal to a combination of coal and biomass has had a significant impact on CO_2 production at the plant:

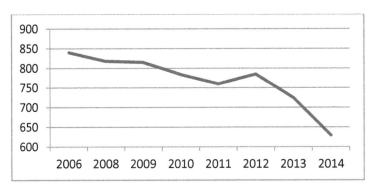

These figures will continue to improve significantly over the next few years as one more generator at the plant switches from coal to biomass and the results of their recently announced carbon capture scheme come online.

None of this work has been cheap, nor has it happened overnight. Yet it proves that old power stations like Drax can be updated and improved throughout their lives.

The argument that this cannot be done economically by the power companies themselves is also disproved by the experiences at Drax. In 2013, Drax Power Ltd posted an annual gross profit of £445m ($707m), based on a turnover of just over £2 billion[52]. Despite significant continuing investment in biofuel processing, interim half-year profits for the first half of 2014 showed a gross profit of £204m on a turnover of £1.25 billion[53]. In the long term, as the increased investment in biofuel processing begins to pay for itself, the company predicts that profits will be significantly higher, thanks to a reduced reliance on imported coal and a reduction in carbon tax.

Gas-fired power stations

Natural-gas-fired power stations are the principal power generators in the UK, after significant investment by the government in the 1980s and 1990s to move away from coal for economic reasons. Until the beginning of the decade, natural gas accounted for around 43% of the country's total energy needs. However, the increasing cost of gas and the drive for renewable energy is now changing this. Gas-fired power is now declining, in the main being replaced by renewable wind power.

Compared to coal, gas-fired power stations have a significantly lower carbon footprint. In addition, the gas itself is usually piped directly to the gas station, which also reduces secondary carbon emissions as there are no transportation costs to get the fuel to the plants.

The switch from coal to gas in the UK was responsible for the significant drop in carbon emissions across the country during the 1990s.

Modern gas-generated electricity has a carbon footprint of around 360g CO_2/kWh (0.88 pounds of CO_2 per kWh), although some of the older gas power stations (using

---

[52]Drax Power Ltd: Annual Report and Accounts for the year ended 31st December 2013.

[53]Drax Power Ltd: Half Year Report for the six months ended 30th June 2014.

'open cycle' technology) have a carbon footprint of around 479g CO_2/kWh (1 pound of CO_2 per kWh).

Oil-fired power stations

Oil-fired power stations are often small power stations that are used to 'top up' the power grid when there is a high demand for power.

There are large oil-fired power stations that are used for generating electricity from oil in the United States.

Oil-fired power stations generate less pollution than coal-fired power stations, but they are not as clean as gas-fired power stations. They also generate more nitrogen oxide, which is regarded as a very significant greenhouse gas, than coal-fired power stations.

Nuclear power stations

Nuclear power stations generate very little air pollution. There do, however, create radioactive waste that has to be stored securely.

Nuclear power stations are extremely expensive to build and to decommission at the end of their lives. They are generally regarded as a low carbon way of generating electricity. However, the environmental impact of building and then decommissioning the plants does pay a big part in the carbon impact of nuclear power. Around one third of the carbon footprint from a nuclear power station over its entire lifetime is created by the decommissioning process[54].

Nuclear power stations generate a constant level of energy and are often used as a base load on the power grids, which is then supplemented by gas- and coal-fired power stations as demand rises.

France uses nuclear power stations to generate the vast majority of its electricity. Many other countries include nuclear in their general power mix.

---

[54]Parliamentary Office of Science and Technology postnote, October 2006 number 268.

Geo-thermal power stations

Geo-thermal energy is extracted from the heat stored deep inside the earth. Boreholes are drilled deep into the earth's surface and water pumped in. Turbines are powered by the super-heated steam that is generated by the heat, and in turn this generates electricity.

Iceland is the pioneer of geo-thermal energy. The country is on the boundary between two tectonic plates in the earth's surface. The high concentration of volcanoes and hot water surface pools have made geo-thermal energy so cheap that in winter some of the pavements in the streets of Iceland's major cities are heated.

Around one quarter of Iceland's electricity comes from geo-thermal energy. In addition, geo-thermal energy provides the heating and hot water needs of almost 90% of the buildings in the country.

Geo-thermal power stations are now being built and tested in different parts of the world, including the United States, the United Kingdom and Canada.

Carbon dioxide emissions from geo-thermal power equate to around 122g CO_2/kWh (around 4½ oz of CO_2 per kWh). Trace amounts of hydrogen sulphide, methane, ammonia, arsenic and mercury are also found and these are captured using emission control systems to filter the exhaust.

Hydro-electric power stations

Hydro-electricity is the most widely used form of 'renewable' energy. Once built, a hydro-electric power station produces no direct waste and has negligible emissions, although there may be environmental issues involved in the construction of the power station in the first place.

20% of the world's electricity is generated by hydro-power and it accounts for around 88% of all electricity from renewable sources.

Most hydro-electric power stations work by damming a river and creating a reservoir. The power comes from the potential energy of the water being driven downhill from the dam under pressure through a water turbine.

Very small-scale hydro-electricity systems may also be powered by a water wheel.

Small-scale hydro-electric systems capable of producing anything from a few kW to around 10 MW (megawatts) of power can be bought off the shelf from various manufacturers and installed at a suitable site within a few weeks. These have become popular in China, where hundreds of 'micro-hydro' power stations have been installed in the past few years.

There are many benefits of hydro-electric power stations:

- There is no 'fuel', so therefore few ongoing costs.

- Emissions are virtually eliminated.

- Hydro-electric power plants have a very long life. Many plants built 80–100 years ago are still in operation today.

- Power production can be increased and decreased at the turn of a tap, which makes them an extremely flexible source of power.

Many hydro-electric power stations have secondary uses. Some are used to manage water flow and stop flooding downstream. Others have become tourist attractions, with the reservoirs being used for water sports and general recreation. In 2009, I visited a hydro-electric power station in India that used the dam for irrigating rice fields around the area with a constant water supply.

Pumped storage

The big issue with electricity production on a huge scale is that it is very difficult to store excess electricity when there is little demand and then use it when demand is at its highest.

This is an increasing problem with the demand for more renewable energy from wind and solar. Wind turbines generate electricity based on the strength of the wind, solar generates electricity based on the strength of the sun. Neither produces electricity on demand when we want it, and this often makes it difficult to incorporate wind and solar energy into our general power mix effectively.

Pumped storage is a way of storing this excess energy and then releasing it on demand when required. The system is based on hydro-electricity, with a reservoir at the top of a hill and a second reservoir at the bottom.

When there is low demand for electricity, such as in the middle of the night, excess electricity is used to pump water from the lower reservoir to the upper reservoir. Water is then released back into the lower reservoir through a water turbine, generating electricity when demand is high.

In effect, the water is being used like a battery, storing electricity until it is needed. Although pumped storage systems are ultimately a net user of electricity, they provide a way of storing excess energy extremely efficiently. Between 70%–85% of the electricity used to pump the water is regained when the water is released.

Like hydro-electric power stations, pumped storage has the huge benefit that it can provide electricity on demand. Quite literally, a pumped storage system can be turned on and off by turning a tap.

Wind turbines

Wind power has been used for thousands of years, operating machinery to grind wheat, pump water or power farming equipment.

Using wind to generate electricity was first tested in Scotland in 1887, and wind turbine production in the United States started the following year. By 1908, there were 72 wind turbine electricity generators in use in the United States, and by the 1930s they were a common sight on remote farms across the country.

Most wind turbines rotate on a horizontal axis and consist of between three and five blades, a gearbox and an electrical generator. The turbine has to face into the wind in order to generate electricity. Small generators are pointed using a wind vane. Large generators use electric motors to move the turbine head into the wind.

Wind turbines are best mounted at height and have to be installed where there is good airflow with minimal turbulence. Large turbines can generate between 2 and 6 MW (megawatts) of power. This is enough to provide power for between 2,000 and 6,000 homes.

Small wind turbines are available that are suitable for installing at homes or businesses. These typically generate between 0.5–6 kW of electricity.

The big issue with wind turbines is they only generate electricity when the wind is blowing, rather than on demand. This is resolved by having lots of wind turbine farms spread over a wide geographical area. Spread wind turbines across an entire country and you are almost always guaranteed that if one area is wind free, another area has strong wind.

The United Kingdom is one of the best locations for wind power in the world. Thanks to significant investment over the past five years, the UK has overtaken France and Italy to become the sixth largest producer of wind power in the world, and it contributes around 8.7% of the UK's electricity demand.

Solar

With a few exceptions, solar has yet to become a serious power generator on a scale to contribute useful amounts of power to a power grid.

There are exceptions. Large solar farms have been built in Canada, Italy, Germany and Spain, the biggest of which is the Sarnia Photovoltaic Power Plant in Ontario, Canada, with almost 100 megawatts of capacity. Other solar farms are being built in Texas, California, Mexico, Iran and Tunisia.

A small amount of electricity can be generated from solar on an overcast day. The majority of the energy production comes on a sunny day. This can be a big benefit when the peak demands for electricity are on a sunny day, such as for running air conditioning units.

Solar is one of the best small-scale electricity generators available. If you require a relatively small amount of electricity at a location and there is no mains power available, solar is often the most cost-effective and easiest energy system to install and run.

Furthermore, quite a few electric car owners have installed their own solar arrays to generate electricity for their electric cars, ensuring that they are truly 'green' vehicles.

'Pit to Plug' – the true carbon footprint for electricity usage

Full lifecycle emissions

In the same way that just measuring the carbon footprint of burning fuel in a car does not show the full well-to-wheel emissions of that fuel, neither does measuring the carbon footprint of burning fuel at a power station show the full well-to-wheel emissions for electricity.

The figures shown earlier in this chapter on emissions from electricity do not include the extraction of the raw materials such as coal or gas. Neither do they include the impact of building the power station in the first place, nor the impact of decommissioning it at the end of its life.

In order to ascertain these figures, we need to look at the full lifecycle emissions for the power station, from the point where it was planned and built, through to its eventual decommissioning.

We also need to take into account losses through the grid: electricity that is lost between the power stations and the home.

Lifecycle emissions for fossil fuel plants

Fossil fuel plants such as coal, oil and natural gas power stations are relatively cheap and simple to build. They last a long time and generate a lot of energy during their lifetimes.

The carbon footprint of a coal- or oil-fired power station is dominated by the emissions during operation. Indirect emissions, such as the raw material extraction and transportation, and the construction of the power station, are relatively minor in comparison.

To put this into context, the pollution caused by mining coal is huge. However, the volume of coal that can be extracted is also huge. As a consequence, the mining itself only adds around 7 g/kWh of CO_2 to the footprint of the coal from an open cast mine, or around 14 g/kWh of CO_2 from a deep underground mine[55]. Rail transportation from the coal pit to a power station adds around 7 g/kWh of CO_2 per kilowatt-hour of energy produced, while international shipping of coal from another part of the world adds more – e.g. shipping from Russia to the United Kingdom, for instance, adds around 2½ g/kWh of CO_2[56].

With gas-fired power stations, the emissions can vary significantly, depending on where the gas originates. The distance from where the gas is sourced to where it is used can make a very big difference to the overall carbon footprint.

If the source is relatively local, such as North Sea gas used in Britain and Norway, for example, the carbon footprint from the extraction and transportation of that gas is around 40 g/kWh. If the source is from further afield, such as Russian or Middle East sourced gas in Britain, the carbon footprint from the extraction and long-distance transportation can be closer to 290 g/kWh[57].

All these figures need to be added to the carbon impact of actually burning the fuel, in order to come up with a true carbon footprint for using gas in electricity generation.

Lifecycle emissions for nuclear plants

Nuclear power stations are very expensive and complicated to build, and expensive to decommission. However, like fossil fuel plants, they last a very long time and generate a lot of energy during their lifetime.

In a nuclear power station, the mining, conversion, enrichment and waste management of the uranium used in the power station accounts for 4–6g of CO_2 per kWh generated, while the actual carbon impact of building the power station in the

---

[55] Carbon Footprint Reduction in Mining and Blasting Operations. Partha Das Sharma, India. October 2009.

[56] Conversations with Drax Power.

[57] Friends of the Earth.

first place (and its decommissioning at the end of its life) equates to around 4g of CO_2 for every kilowatt-hour of electricity produced.[58]

Lifecycle emissions for renewable energy

The carbon footprint for renewable energy sources (solar, wind, hydro) is dominated by their emissions during manufacture and their decommissioning costs when they reach the end of life.[59]

Solar and wind generators have a shorter lifespan than fossil fuel plants. However, because renewable energy sources generate no emissions during use, they are still very carbon efficient.

For instance, solar power itself may be entirely carbon neutral, but when you take into account the actual production of the solar panel, the carbon footprint is in the region of 35–60 g/kWh, depending on where you live and how much sun you have.

Large wind turbines are still very carbon efficient. Large wind turbines built on land typically have a carbon footprint of around 4½ g/kWh generated during their lifetime, while wind turbines erected at sea typically have a carbon footprint of around 5½ g/kWh.

Hydro-electric power stations have the longest lifespan of any power station. Many of the hydro-electric power stations in Scotland were built in the 1930s and are still generating large amounts of power today.

Hydro-electric power stations built on fast-running rivers can have a carbon footprint of around 5 g/kWh, while the environmental impact of building huge dammed reservoirs for hydro-power can have a carbon footprint of up to 32 g/kWh over the lifetime of the power station.

---

[58] British Energy – Carbon footprint of the nuclear fuel cycle: briefing note. March 2006.

[59] Parliamentary Office of Science and Technology: Postnote. October 2006, number 268.

Utility grid transmission losses

As electricity is transmitted from the power stations to the consumers, a certain amount of energy is lost on route. These losses are reduced by transmitting electricity at very high voltages. In the US, 7.2% of all energy generated is lost through utility grid transmission[60]. In the United Kingdom the figure is 7%[61].

From pit to plug: the true carbon footprint of electricity

Now that we know the full impact of using electricity, we can put together a more accurate carbon footprint for using electricity to power an electric car:

	CO_2 emissions per kWh
Coal	1077g (2 pounds 6½ ounces)
Gas	470–738g (between 1 pound 1 ounce and 1 pound 10 ounces)
Oil	802g (1 pound 12½ ounces)
Nuclear	22.5g (less than 1 ounce)
Geo-thermal	150g (5¼ ounces)
Hydro-electric	5–32g (1 ounce)
Wind turbine	4.5–5.5g (¼ ounce)
Solar	35–60g (1¼ – 2 ounces)

---

[60] US Climate Change Technology Program – Technology Options for the Near and Long Term, November 2003.

[61] Investigation into Transmission Losses on UK Electricity Transmission System. National Grid Technical Report. June 2008. / Electricity Distribution Losses – a Consultancy Document. Ofgen. January 2003.

These figures include the sourcing and transportation of the fuels, consuming the fuels and, with the exception of solar where the electricity is usually used close to the point of generation, the transmission losses between the power station and the consumer.

The figures also take into account the construction of the power station in the first place and the decommissioning of the power station at the end of its life.

A more accurate electricity carbon footprint by country

On page 140, I produced a table of carbon emissions from energy generation broken down by country. These figures were purely based on the emissions from the power stations rather than on the true carbon footprint of electricity at the plug. However, using these figures as a starting point, it is possible to provide an estimate of the true carbon footprint once we take all the other parameters into account.

At this point, I have to say I was unable to find a set of multi-national figures that I could rely on. So the figures in the table below are my own calculations, based on getting the best information available. This has been done by establishing the mix of different power sources by percentage for each country and weighting them accordingly.

Country	CO_2 emissions per kWh
Austria	261g (9 ounces)
Australia	922g (2 pounds 1 ounce)
Canada	258g (9 ounces)
Denmark	380g (13½ ounces)
France	90g (3½ ounces)
Finland	230g (8 ounces)
Germany	679g (1 pound 8 ounces)

Greece	865g (1 pound 15 ounces)
Iceland	163g (6 ounces)
India	954g (2 pounds 2 ounces)
Ireland	701g (1 pound 9 ounces)
Italy	593g (1 pound 5 ounces)
Japan	590g (1 pound 5 ounces)
Nepal	77g (3 ounces)
New Zealand	322g (12 ounces)
Norway	76g (3 ounces)
Pakistan	456g (1 pound)
Spain	518g (1 pound 3 ounces)
Sweden	68g (2 ounces)
United Kingdom	620g (1 pound 6 ounces)
United States of America	697g (1 pounds 8½ ounces)

As a basis for serious scientific research, these figures should not be relied upon, although I am happy to provide the details of my calculations on request. However, they do provide a useful benchmark for our purpose of establishing a base figure for the environmental benefit of electric cars.

You will notice there are some significant differences between the figures shown here and the figures shown on page 140. Of course, all the figures are higher, but at first glance many of the figures seem to be disproportionally high. Proportionally speaking, Finland, Norway and Nepal have very much bigger carbon footprints (although they

are still very low overall), while the increased carbon footprint in the United Kingdom, Germany and Italy are also higher than average.

In the case of Finland, Norway and Nepal, the increase is due to the fact that, in official figures, renewable energy is regarded as zero emission energy. My estimated figures take into account the construction, maintenance and eventual decommissioning of the hydro-electric power stations to give a more realistic figure.

In the case of the United Kingdom, Germany and Italy, a substantial amount of the difference is down to the carbon cost of transporting gas to the power stations. Piped gas from the North Sea is comparatively low carbon, while gas from Russia and liquefied gas from the Middle East has a significantly higher carbon footprint. Whilst the United Kingdom's demand for gas has dropped over the past four years, the country is now a net importer of gas and oil, which thereby increases the carbon footprint for these products.

Electric cars and electricity supply

Critics claim that the power grid cannot provide enough electricity for electric cars without significant and costly upgrading.

It is true that, if every combustion engine car was taken off the road and replaced with an electric one, electricity consumption would increase. It has been estimated that across the European Union, net electricity consumption would increase by 15%[62]. Worldwide, a total transport switch to electric vehicles would increase electricity consumption by 20%[63]. However, a complete 100% shift to electric power is extremely unlikely. Neither governments nor car manufacturers are anticipating a complete switch at any time over the next 40 years. Indeed, most predictions show that even by 2050, only around 60% of the cars on the road will be electric-powered.

Electricity companies and governments have investigated the likely power demand based on a number of different scenarios for electric car take-up. The results make interesting reading:

- In Germany, one million electric cars travelling an average of 10,000 km a year would require less than 1% of its current electricity capacity in order to provide sufficient energy[64].

---

[62]The Future of Transport in Europe: Electricity Drives Cleaner! Eurelectric (2009).

[63]Energy Technology Perspectives 2008: In Support of the G8 Plan of Action. Scenarios and Strategies to 2050. IEA (2008).

[64]The Electrification Approach to Urban Mobility and Transport. Strategy Paper. ERTRAC. 24th January 2009.

- Likewise, the UK Department for Transport claims that the UK has sufficient generating capacity to cope with the uptake of electric cars, assuming a managed charging-cycle targeted at off-peak periods (particularly at night) when there is surplus capacity[65].

The study carried out by the Department for Transport does suggest that if significant numbers of owners started charging their cars during peak hours, significant investment may be required in the longer term.

- In the United States, the American grid could support 94 million electric vehicles (43% of all cars on the road) if they were all charged during the evening and overnight, or 158 million vehicles (73% of all cars on the road) with the more advanced charging techniques currently being developed[66].

It is widely expected that the majority of electric cars will be charged up overnight. This coincides with the time when there is surplus capacity for electricity production. This will negate the need for investment in power upgrades. Power companies are already promoting electric car night-time charging using smart metering (see page 170) and discounted energy tariffs.

In the course of writing this book, I have spoken at length with power network infrastructure specialists in the United Kingdom, California, India and France about the impact of electric cars on the power grid. There is widespread agreement that even a significant take-up of electric cars is unlikely to cause problems in the next ten years. In the longer term, there is an expectation that a substantial take-up of electric cars and rapid charging infrastructure may contribute to increased peak demand for electricity by 2025–2030.

---

[65]Investigation into the scope for the transport sector to switch to electric vehicles and plug-in hybrid vehicles – United Kingdom Department for Transport (2008).

[66]Electric Powertrains: Opportunities and Challenges in the US Light-duty Vehicle Fleet. Kromer and Heywood, Laboratory for Energy and the Environment, Massachusetts Institute of Technology.

Smart metering

Smart meters are the next generation of electricity and gas meters, providing customers and energy suppliers with accurate information about the amount of energy being consumed at any one time and the cost of that energy.

Smart meters also allow energy suppliers to provide flexible tariffs to their customers so that electricity costs can be cheaper when demand is low and higher when demand is high. This offers consumers the choice of when they use their energy and encourages them to use energy as efficiently and as cost-effectively as possible. Many advanced smart-metering systems can also be configured to switch appliances on and off, depending on the cost of the energy.

In order to reduce the impact of electric cars on the power grid over the next few years, it is likely that electric car owners will be encouraged to install smart meters that automatically switch the cars on to charge when demand for electricity is low and therefore cheap. Owners will be encouraged to charge their cars at night-time with lower cost electricity, thereby reducing the impact of a large take-up of electric cars.

Monitoring the emissions of your own electric car

In the meantime, it is possible for you to monitor the emissions from your own electric car by monitoring the supply and demand on the power grid yourself.

The *www.TheElectricCarGuide.net* website includes a data feed from the UK National Grid showing the power supply mix and average carbon footprint of the electricity being generated across the UK. This information is updated every five minutes throughout the day. The site also recommends the best times of day to plug in your electric car, based on electricity demand and carbon footprint.

Fuel economy

Finally, we now have a way of comparing the environmental performance of an electric car to a conventional car running on petrol, gasoline or diesel. Using the information we now have, it is possible to measure the CO_2 emissions of an electric car and compare them with the CO_2 emissions of a combustion engine car.

Depending on the make and model, most electric cars can travel between 3½ and 7½ miles (6–12 km) on a single kilowatt-hour of electricity.

In the chapter *Electric Cars You Can Drive Today* (starting on page 78), I include a list of all the electric cars that are available, along with the manufacturer-supplied fuel economy figures, expressed in the number of km you can travel on one kilowatt-hour of electricity (km/kWh).

Based on the carbon footprint of the electricity used to charge up the cars, we can therefore provide a *well-to-wheel* carbon footprint for running an electric car.

In the European Union, emission figures for cars are shown as the number of grams of CO_2 generated in a single km of driving (g CO_2/km). In the USA, the Environmental Protection Agency has similar calculations, based on grams of CO_2 per mile.

For these reasons, I am showing these comparative figures in metric measurements. If you want to work in pounds and ounces, one ounce is 28g and one pound is 454g.

	Grams of CO_2 per km						
Kilometres	6km	7km	8km	9km	10km	11km	12km
Miles	3.75m	4.38m	5m	5.62m	6.2m	6.9m	7.5m
Austria 265g/kWh	44g	38g	33g	29g	27g	24g	22g
Australia 948g/kWh	160g	137g	120g	106g	96g	87g	80g
Canada 281g/kWh	47g	40g	35g	31g	28g	26g	23g
Denmark 370g/kWh	62g	53g	46g	41g	37g	34g	31g
France 100g/kWh	17g	14g	13g	11g	10g	9g	8g

Country							
Germany 626g/kWh	104g	89g	78g	70g	63g	57g	52g
Iceland 163g/kWh	27g	23g	20g	18g	16g	15g	14g
India 952g/kWh	159g	136g	119g	106g	95g	87g	79g
Ireland 753g/kWh	126g	108g	94g	84g	75g	68g	63g
Japan 530g/kWh	105g	90g	79g	70g	63g	57g	53g
Nepal 77g/kWh	13g	11g	10g	9g	8g	7g	6g
N. Zealand 330g/kWh	55g	47g	41g	37g	33g	30g	28g
Norway 76g/kWh	13g	11g	10g	9g	8g	7g	6g
Sweden 68g/kWh	11g	10g	9g	8g	7g	6g	6g
UK 650g/kWh	108g	93g	81g	72g	65g	59g	54g
USA 731g/kWh	122g	104g	91g	81g	73g	66g	61g

As you can see, there is a huge difference in CO_2 emissions, depending on which country you live in. If you live in Norway, Sweden or Nepal, where most of the electricity is generated with hydro-power, your CO_2 emissions are very low indeed if you drive an electric car.

In countries such as Australia and India, this chart suggests there is only a small benefit of an electric car over an equivalent combustion engine car. In reality, there is a saving once well-to-wheel emissions for combustion engine cars are taken into account, and I look at this in more detail on the next page. I also demonstrate this difference in my chapter on real world economy tests, starting on page 183.

As there is wild variation between CO_2 emissions in each country, it is not possible to provide a blanket CO_2 per km measurement. It is, however, possible to create this measurement by country:

Make and Model	Canada CO_2	France CO_2	India CO_2	UK CO_2	USA CO_2
BMW i3	26g/km	9g/km	87g/km	59g/km	66g/km
Mitsubishi i-MiEV	28g/km	10g/km	95g/km	65g/km	73g/km
Nissan LEAF	35g/km	13g/km	119g/km	81g/km	91g/km
Renault Zoe ZE	31g/km	11g/km	106g/km	72g/km	81g/km

How to further improve an electric car's CO_2 footprint

In countries where there is a comprehensive mix of different power generation sources, high carbon sources of electricity production such as coal-fired power stations are often put on standby when electricity demand is low. Instead, electricity is typically generated by nuclear and hydro-power stations.

In the UK, where the national power grid provides detailed power information every five minutes throughout the day, it is possible to identify at any time how much power is being produced and what the power source is.

This information is published and available for public inspection. If you want to see the current carbon footprint of the electricity used in the UK, you can monitor the UK power grid by visiting *www.TheElectricCarGuide.net* and following the 'How green is my power?' links.

Based on this information, you can quickly identify the best time to charge your electric car in order to reduce your carbon footprint. In the UK, the carbon footprint for a kilowatt-hour of electricity varies from around 330–420 g/kWh during the night, up to 660–725 g/kWh during the late afternoon and early evening.

In almost every country in the world, the carbon footprint of the electricity supply falls between 11 pm and 7 am. If you can charge up your car between those times, your carbon footprint for using your electric car will be lower than if you charge up at any other time.

Carbon footprints for charging an electric car at different times in the UK

Make and Model	UK Average (650g/kWh)	UK overnight (380g/kWh)	UK peak – early evening (700g/kWh)
BMW i3	59g/km	35g/km	64g/km
Mitsubishi i-MiEV	65g/km	38g/km	70g/km
Nissan LEAF	81g/km	48g/km	88g/km
Renault Zoe ZE	72g/km	42g/km	78g/km

As you see, there can be a big difference in the carbon footprint of an electric car, depending on when it is charged.

Fuel economy for combustion engine cars

By way of comparison, here are published manufacturers' figures for some of the most efficient petrol/gasoline and diesel-powered cars currently on sale today, shown as 'tank-to-wheel' CO_2 figures.

Alongside them, I have shown an estimated 'well-to-wheel' figure based on the CO_2 impact of producing the fuel in the first place (based on the figures and calculations shown on page 146) by multiplying the tank-to-wheel figure by 1.205.

Make and Model	Tank-to-wheel CO_2 figures	Well-to-wheel CO_2 figures
BMW 116i	131g/km	158g/km
Chevrolet Aveo	119g/km	143g/km
Fiat 500 Twin Air	95g/km	114g/km
Ford Fiesta ECOtec (diesel)	98g/km	118g/km
Ford Focus 1.0 Ecoboost	109g/km	131g/km
MINI One Diesel	89g/km	107g/km
Range Rover Evoque 150bhp	133g/km	160g/km
Smart ForTwo diesel	88g/km	106g/km
Smart ForTwo mhd	100g/km	122g/km
Toyota Auris hybrid	91g/km	110g/km
Toyota Aygo	95g/km	114g/km
Toyota IQ 1.0 VVTi	99g/km	119g/km
Toyota Prius	89g/km	107g/km
Volkswagen Polo Bluemotion	99g/km	119g/km
Volvo V30 1.6D DRIVe	99g/km	119g/km

It is interesting to see how these figures have fallen since I first compiled this list in 2010. Car manufacturers have made great strides forward in the past few years. Yet if you compare these figures with the electric car figures shown on page 173, you will see that electric cars still provide significantly lower well-to-wheel emissions than combustion engine cars in almost every country in the world.

Real world testing

So far, I have been using manufacturer's figures for car economy. Real-world economy can vary significantly from manufacturer's figures.

In order to provide a far better comparison, I undertook my own tests to measure real-world carbon emissions of both electric cars and combustion engine cars, driving on a variety of roads.

You can read the results of these tests, starting on page 183.

The environmental impact of batteries

It is important to measure the environmental impact of batteries (both their construction and what happens at the end of their life) when considering an electric car.

There are two main battery technologies that are now used in electric cars. Lead acid batteries are used in smaller, lower-powered vehicles. Larger and more powerful vehicles are now using lithium ion batteries.

Nickel metal hydride batteries (as used in the Toyota Prius and the Honda Insight) are now rarely used in full electric vehicles.

Lead acid batteries will need to be replaced several times during the life of the vehicle. Depending on the exact technology used, lithium ion batteries may require replacement once during the lifetime of the car, although some manufacturers, like Mitsubishi, expect the batteries on their electric cars to last the lifetime of the vehicle.

In order to provide a comparison between the different batteries, I am going to assume that a car has a lifespan of twelve years. If the car uses lead acid batteries, I am going to assume that during this time it will drive 50,000 miles (80,000 km). If the car uses lithium ion batteries, I am going to assume that during the car lifetime, it will drive 100,000 miles (160,000 km).

I have assumed two different distances because of the different characteristics of lead acid and lithium ion batteries:

- Lead acid batteries are typically used in small, speed-restricted electric vehicles that are more appropriate for town or city use. They have a comparatively short range and are therefore used by people who do not travel very far.
- Lithium ion batteries are much lighter and more powerful and are typically used in faster and longer-range electric cars.

We can then use these figures to come up with an approximate CO_2/km figure for the environmental impact of the batteries used in an electric car.

Lead acid batteries

Until very recently, lead acid batteries were the mainstay of the electric car industry. They are still used by most neigborhood electric vehicles (NEVs) and some quadricycles in Europe. It is easy to see why they are popular for lower cost vehicles. They provide a reliable source of power at a comparatively low cost and they are simple batteries to integrate into a vehicle.

The disadvantage of lead acid batteries is that they have a typical lifespan of only three to five years. At the end of their life they need to be disposed of.

A typical electric car equipped with lead acid batteries will have 8–12 batteries, providing a total of 8–12 kWh of energy storage.

Lead acid batteries are a simple technology that is cheap and easy to manufacture. Their cases are typically made of polypropylene, the plates are made of lead, and a mixture of acid and water is used as an electrolyte.

Many lead acid batteries also have a very high percentage of reused materials: both the case and the plates are typically manufactured by recycling old lead acid batteries.

Consequently, the carbon cost of manufacture is low. Around 15 kg of CO_2 is generated through the build of a typical 1 kWh lead acid battery.[67]

Thanks to the value of the raw materials, there is an active market in recycling lead acid batteries at the end of their lives. In Europe, virtually 100% of lead acid batteries are recycled. In the United States, the figure is around 98%.

97% of a lead acid battery can be recycled. The casing can be melted down and reformed. Lead has a low melting point and, as a consequence, it requires very little energy to be converted back into a raw material and reused.[68]

Lead acid batteries are typically recycled to produce new lead acid batteries, thereby keeping the carbon footprint down for the next generation of batteries.

Based on an electric vehicle requiring eight lead acid batteries which are then replaced every four years, the carbon footprint for the batteries in an electric car, spread over the lifetime of the vehicle, is as follows:

Estimated carbon footprint per battery:	15 kg
Estimated carbon footprint per battery pack (8 batteries):	120 kg
Number of battery packs required during the lifetime of the vehicle:	3
Carbon footprint over lifetime of vehicle:	**360 kg**
CO_2/km based on 50,000 miles (80,000 km)	4½ g

---

[67]Source: HGI: How Green is a Lead Acid Battery? A study of the environmental impact of the lead acid battery.

[68]Source: International Lead Association: fact sheet on lead recycling.

Examples of vehicles with this configuration include the REVA *i* (G-Wiz), the Aixam Mega City and most NEVs in the United States.

Lithium ion

Lithium ion batteries have a significantly longer lifespan than lead acid batteries and are designed to last the lifetime of the vehicle.

The carbon cost for manufacturing a typical lithium ion battery pack for a car is around 22 kg for every 1 kWh of energy storage[69]. Most small lithium ion equipped electric cars have 16–24 kWh of batteries. This gives an equivalent carbon footprint of between 352 kg and 528 kg per vehicle.

When lithium ion batteries reach the end of their useful lives, the metals from the batteries can be extracted. At present, around 85% of the materials from the batteries can be recovered and reused for making new batteries. By the end of the decade, it is anticipated that these figures will be up to 97%.

It has been estimated that new lithium ion batteries produced from recovered lithium will have a manufacturing carbon footprint of around 6.8 kg per kWh of energy storage.[70]

Estimated carbon footprint for a 16kWh battery pack:	352 kg
CO_2/km based on 100,000 miles (160,000km)	2.2 g

The environmental impact of vehicle manufacturing and distribution

In the United Kingdom, the Society for Motor Manufacturers and Traders (SMMT) estimated in 2005 that the energy needed to manufacture a car in the United Kingdom translated to 600 kg of CO_2, down from an estimated 1,100kg in 1999. In

---

[69]Source: Umicore, based on the CO_2 emissions of Saft Li-Ion MP 176065 cells.

[70]Source: Umicore: Strategic choices in Li-Ion and NiMH battery recycling.

addition to these figures, the SMMT estimated that the production of the raw materials added a further 450 kg per vehicle, although many environmental experts believe this raw material carbon estimate is much too low.

In the United States, Honda estimates their average to be 810 kg of CO_2. At the time of writing, they do not have any estimates for the raw material impact.

More recently, Kia carried out a total lifecycle analysis assessment of their new Cadenza mid-range car, measuring the carbon footprint of the car from the point of manufacture to the point of destruction. The work was certified by the Korea Environmental Industry and Technology Institute (KEITI).

According to their figures, production of the raw materials for the Kia Cadenza causes the emission of 3.48 tons of carbon dioxide. A further 0.531 tons are produced in the production of the car and 0.012 tons during recycling[71]. These figures are dwarfed by the carbon dioxide emissions from actually using the car: 25.5 tons of carbon dioxide emissions based on 75,000 miles (120,000 km) of driving.

At present, no vehicle manufacturer currently publishes carbon footprint figures for the production of an electric car. I have discussed footprint figures with a number of manufacturers. I have been told that the carbon footprint involved in the production of an electric car is lower than that of a conventional car because of the fewer mechanical components in the vehicle.

Electric vehicle manufacturer Mahindra REVA believes its carbon footprint for an electric car is significantly lower than the carbon footprint for other vehicles. Mahindra REVA claims that its electric cars use around 80% fewer parts than conventional cars, ensuring that they have a very low carbon footprint.

---

[71]Source: Hyun-jin Cho, Sustainability Manager, Kia Motors.

International shipping

In terms of environmental impact, it makes very little difference whether your car was shipped around the world from the factory to the customer or whether it was built a few miles down the road. Less than 1% of the overall carbon footprint of the average vehicle is associated with distribution[72]. International shipping makes up the smallest part of the carbon footprint associated with distribution.

A much greater part of the footprint associated with vehicle distribution is attributed to the road transportation of the car.

Vehicle recycling

Car manufacturers and governments alike have worked hard to ensure that when a car comes to the end of its life, as much of it as possible can be recycled.

Most car manufacturers already go further. Since the early 1980s, Mercedes-Benz has been using recycled materials in the production of its cars, and today most cars have a proportion of recycled materials in their build.

Across Europe, every post-1980 car that is scrapped at the end of its life must be recycled, with a minimum of 85% of its content reused.

In Europe, car manufacturers are now legally responsible for ensuring that cars they manufactured are properly recycled at the end of their lives. In the vast majority of cases, well over 90% of the content of a scrapped car is now reused through vehicle recycling.

It is too early to say whether an electric car is more or less recyclable than any other type of car. It is unlikely that it will be much different from existing vehicle recycling. Battery recycling is already well established and the high value of the metals used in

---

[72]Ecolane Transport Consultancy: Life Cycle Assessment of Vehicle Fuels and Technologies – March 2006.

both the batteries and the motor in an electric car will ensure it is financially profitable for vehicle recyclers to dismantle electric cars properly.

Chapter summary

Buying and running an electric car has significant environmental benefits over any other type of car:

- Electric cars emit no pollution at the point of use, thereby reducing air pollution in our towns and cities.

- The pollution emitted from power stations in order to charge up an electric car is less than the pollution emitted from the exhaust pipe of a conventional car.

- The carbon footprint associated with electric car production is claimed to be lower than other vehicles, although there is currently no hard data to back this up.

- The carbon footprint associated with driving an electric car is less than that of other vehicles, even when battery replacement and recycling is taken into account.

- The issues that used to exist with recycling lithium batteries have been addressed and are improving year on year.

- By the time electric cars that are built now are being recycled, it is expected that 95-97% of all lithium will be recoverable from a lithium battery.

Real World Economy Figures for Electric Cars

In the real world, with real-world driving, it seems virtually impossible to get the fuel economy figures that the manufacturers claim for their models. Whether you are talking about combustion engine cars or electric cars, the figures appear to have little relevance in the real world other than to work as a comparison with other makes and models.

In an electric car, however, you can create your own fuel economy measurements extremely easily. Simply record your distance travelled and then measure the amount of electricity used to recharge the car using a plug-in watt-meter.

The test

In order to measure real world economy for electric cars, I carried out a driving test along a fixed route with two different electric cars.

In order to provide a useful comparison, I then tested two combustion engine cars, driving the same route, in order to identify real-world fuel economy and comparative carbon footprint figures for each type of vehicle.

I decided to use my own personal commute to and from work as a test route, travelling in busy traffic. The distance travelled is 7 miles (approximately 11¼ km) each way. The route comprises 2½ miles (4 km) of fast freeways and 4½ miles (7¼ km) of busy inner-city roads.

The tests were carried out in and around Coventry in the United Kingdom in January 2010. Temperatures were around freezing during the whole trial. Cabin heating was used as appropriate.

Noting the temperature is important. Electric cars are less economical in cold conditions than they are in hot conditions. These tests therefore reflect a 'worse case' economy for electric cars. In warmer conditions, it would be fair to expect significantly improved figures on an electric car.

Likewise in cold conditions, combustion engine cars take longer to warm up and are also not at their most economical at the start of their journey.

The carbon calculations

The results of these tests were originally published in my book, *Owning an Electric Car*, published in February 2010. The economy figures for the electric cars published in that book did not fully take into account all the emissions from generating electricity. Specifically, the calculation overlooked the following:

- The sourcing of the fuel
- The carbon cost of building the power station in the first place
- The carbon cost of running the power station
- The carbon cost of decommissioning the power station at the end of its life

These test results have now been updated to reflect these additional carbon figures. It is worth noting that adding these additional carbon figures makes very little difference to the carbon footprint of an electric car, and the conclusions from the original tests remain unchanged in light of these updated figures.

Test validity

It is worth stressing that these tests have not been independently verified by any scientific establishment. Consequently, these tests can only ever be used as an indication of relative fuel economy and carbon emissions.

I also feel that it is important that my test could be repeated by anyone else using their own cars and their own routes, and the tests have been simplified in order to achieve this.

All the information and calculations I used in order to carry out my tests are included within this chapter. If a university or a scientific establishment wishes to carry out similar tests in a controlled environment and would like to discuss my test methods, I can be contacted through the *Ask Me a Question* page of the website *www.TheElectricCarGuide.net*.

The electric cars

Two electric cars were chosen: a brand new Mitsubishi i-MiEV and a three-year-old REVA G-Wiz dc-drive with old batteries.

The Mitsubishi was chosen as an excellent example of the latest technology electric car. It is a sub-compact car with a roomy interior, providing good performance and range.

The REVA dc-drive was chosen to identify whether an electric vehicle remains environmentally efficient as the batteries degrade and the car gets older. The REVA dc-drive is no longer on sale; it has been replaced by the more efficient REVA *i* and REVA L-ion. It represents a good test on an older electric car.

At midnight each night, the cars were plugged in until the battery packs were completely charged. The amount of electricity used was monitored using a watt-meter and this was then multiplied by the average carbon footprint for UK electricity during the period the cars were on charge. I was charging the cars overnight, when the power grids are under-used, and the carbon footprint averaged out at 380 g/kWh.

This carbon footprint figure takes into account the carbon impact of sourcing the fuel and transporting it to the power station, the production of the power and the average transmission losses of the power as it is delivered from the power station to the car.

You can view these figures yourself on the *How Green is the Grid?* web page on *www.TheElectricCarGuide.net*.

I also recalculated the carbon footprint figures based on the environmental footprint of a coal-fired power station, plus the mining and transportation of the coal. Friends of the Earth estimate that this figure is 1007 g/kWh. I then added 7% to this figure to take into account energy losses between the power station and the home, giving a total of 1077 g/kWh.

Finally, I took into account a carbon footprint for the use of the batteries, using the calculations shown on page 176.

The combustion engine cars

The combustion engine cars chosen were a brand new Toyota Aygo 1.0 and a Fiat Panda 1.1.

Both of these cars are economical sub-compact city cars that produce low levels of carbon emissions. The manufacturers' own CO_2 footprint figures show that, in official tests, the Toyota Aygo produces a tank-to-wheel footprint of 106 g CO_2/km, while the Fiat Panda produces a tank-to-wheel footprint of 119 g CO_2/km.

These figures only reflect the tank-to-wheel emissions, not the well-to-wheel emissions. For our tests to be comparative to the electric car tests, well-to-wheel calculations have to be used.

In order to calculate the carbon footprint in real conditions, I filled the fuel tank at the start of the test. I then measured the fuel economy in litres at the end of each test by refilling the fuel tank. I calculated the CO_2 footprint based on the amount of fuel used, using the 'well-to-wheel' CO_2 figures shown on page 146.

Test results from the electric cars:

	i-MiEV	GWiz
Distance travelled	14.1 miles 22.56 km	14.1 miles 22.56 km
Electricity used	3.19 kWh	2.99 kWh
Total electricity cost[73]	26p UK 31¢ US	24p UK 30¢ US
Average CO_2 per kWh	380g/kWh	380g/kWh
CO_2/km electricity usage	46.66 g/km	43.73 g/km
CO_2/km battery usage[74]	2g/km	6g/km
Total CO_2/km	52.73 g/km	50.36 g/km

---

[73] Costs based on a night time tariff of 8p per unit in the UK, and 10¢ per unit in the United States.

[74] See the section on the Environmental Impact on Batteries on page 130 for a definition for this figure.

These figures show remarkable fuel economies and a low recharge cost for the electric cars. The carbon footprint is also low, which is helped by using off-peak electricity. Off-peak electricity is much more carbon friendly than using electricity during peak times.

What if they were powered by coal?

When electric cars are powered by coal-fired power stations, the carbon footprint is significantly higher than when they are powered by most other sources.

Using the figures of 1077g/kWh for coal-fired power, including the mining and transportation of the coal and the energy losses between the power station and the home; this is what the CO_2/km would look like if I charged up using coal power:

	i-MiEV	GWiz
CO_2/km electricity usage based on coal power	151.28 g/km	142.74 g/km

How little electricity an electric car uses

To put these figures into context, I recently measured the amount of energy used by a Bosch tumble dryer. To dry a single load of clothing used 2.97 kWh of electricity – almost exactly the same amount of energy used by the REVA G-Wiz in these tests, and only marginally less than the Mitsubishi i-MiEV.

A family of four can quite easily wash and dry four loads of clothes per week. That is the equivalent of around 56 miles (90 km) of driving every single week, or a total of almost 3,000 miles – 4,800 km – of driving each year.

If clothes could be dried outside on a washing line or a clothes airer for even six months every year, the energy saving would be enough to drive a considerable distance using the electricity saved.

Test results from the combustion engine cars

	Toyota Aygo	Fiat Panda
Distance travelled	14.1 miles 22.56 km	14.1 miles 22.56 km
Fuel used	1.33 litres	1.51 litres
Total fuel cost[75]	£1.71 UK $1.04 US	£1.94 UK $1.15 US
Official CO_2/km tank-to-wheel	106g/km	119g/km
Actual CO_2/km tank-to-wheel	136.48 g/km	154.94 g/km
CO_2/km well-to-wheel	164.46 g/km	186.70 g/km

As you can see, the carbon footprint figures that I achieved in my test are significantly higher than the official CO_2 figures. There are various reasons for this:

- The cars were driven by me and not a professional test driver. While I did use eco-driving techniques, I would never claim to be the best eco-driver in the world!
- The tests started with cold engines in cold conditions.

---

[75] Fuel price based on a UK cost of £1.14 per litre and a US cost of $2.60 per US gallon.

- The cars were driven on a variety of roads, including freeways, suburban roads and city streets in heavy traffic conditions.
- Cabin heating was used in the cars as appropriate to keep the windscreen clear (this is also true of the electric cars).

Side by side analysis: well-to-wheel measurements

	Mitsubishi i-MiEV	REVA GWiz	Toyota Aygo	Fiat Panda
Fuel cost	26p UK 31¢ US	24p UK 30¢ US	£1.71 UK $1.04 US	£1.94 UK $1.15 US
CO_2/km	52.73 g/km	50.36 g/km	164.46 g/km	186.70 g/km

Side by side analysis if powered by coal

If the electric cars had been powered by a typical coal-fired power station, the carbon footprint comparison would look like this:

	Mitsubishi i-MiEV	REVA GWiz	Toyota Aygo	Fiat Panda
CO_2/km	152.28 g/km	142.74 g/km	164.46 g/km	186.70 g/km

Chapter summary

- I have carried out tests on the comparative economies of two different electric cars and two economical combustion engine cars.
- As these tests have not been independently verified, they can only ever be used as an indication of relative fuel economy and carbon emissions.

- The tests indicate that electric cars are significantly better for the environment than the equivalent combustion engine car.
- Even if my electricity is generated by a coal-fired power station, my tests indicate that an electric car is still better for the environment than a combustion engine car.

A Final Word

Electric cars are radically different and an exciting new technology that has practical uses in our daily lives. They have significant environmental and economic benefits.

For many people they are the ideal vehicle, providing quiet, smooth and practical motoring.

If I have encouraged you to go out and try an electric car or, even better, given you the confidence to get one for yourself, I have achieved what I set out to do in writing this book.

Likewise, if you have read the book and come to the conclusion that an electric car is not the right choice for you, this book has also served its purpose. Far better to spend a small amount of money buying a book than spend a lot of money on buying the wrong car.

The electric car industry is a very exciting one and over the next few years there will be some very significant new developments. For this reason, I will be producing regular updates to this book in order to keep up with the latest technology, ideas and products.

Finally, if you have enjoyed this book, or even if you haven't, feel free to get in touch. If you have questions about electric cars, or suggestions on how I can improve the book, I would be delighted to hear from you. I can be contacted through the 'Ask Me a Question' page on my website *www.TheElectricCarGuide.net*.

All the best,

Michael Boxwell
April 2016

Appendix A: Electric Vehicles in Business

Electric vehicles can make a lot of sense within a business environment. They provide economical transportation for inner-city use and generate significant interest from both customers and the surrounding community, creating good publicity and goodwill.

However, they are not suitable for all businesses. Some businesses have had a lot of success using electric vehicles, while others have experienced issues and found themselves running their businesses around the electric vehicles.

If you are planning to use an electric vehicle for business, you need to make sure they are going to be a success. For that reason, it is important to carefully evaluate whether or not an electric vehicle will suit your business needs.

Types of electric vehicle available

Electric Cars

Many businesses have one or more cars available for staff to visit customers. Estate agents, accountants, legal companies and insurance agents often have members of staff who need to be able to visit client sites, sometimes at a moment's notice.

Some businesses have found that an electric car is ideally suited to this use. If you are based in a city and all your clients live in the same city, an electric car can be a quick way to get around.

Thanks to their ease of use, different members of staff can quickly get used to driving the cars, and electric cars have proven popular with employees and customers alike.

If your business operates from a campus, a low-powered electric car or NEV can be a great vehicle for personnel movements and internal postal deliveries. The United States Army uses several thousand GEM electric NEVs specifically for this purpose.

Electric minibuses and buses

Electric minibuses and buses are available. These range from 6- to 36-seat models. These are normally used for infrequent shuttle service rather than constant use, usually on large campuses, by hotels shuttling customers to and from airports or by retailers shuttling customers to and from their stores.

Commercial vehicles

If you use vans, trucks or heavy goods vehicles for mainly city area and urban area work, there are now a number of electric vehicles available.

Many of the electric vans and trucks have a limited top speed, typically ranging from between 30 mph (45 km/h) and 50 mph (80 km/h), depending on the make and model. This limitation is rarely a problem in built-up areas, where the high torque from the motor ensures that the vehicles can keep up with other traffic on the roads, but it may be a limitation if using an electric vehicle outside these areas.

This is not the case with the Nissan e-NV200 and Renault Kangoo ZE electric vans. These are both highway-capable vehicles that are capable of high-speed cruising as well as shorter journeys. As with all electric vehicles, range does drop significantly if driving constantly at high speeds, and this is more pronounced in a commercial vehicle, partly because of its bigger and bulkier shape, but also because commercial vehicles are usually carrying a fairly heavy load, which also impacts on range.

Electric bicycles, motorbikes and scooters

If you use motorbikes for delivering small packages or take-away food to the surrounding area, an electric bicycle or electric motorbike may be a suitable option.

Insurance for pizza delivery motorbikes is expensive, whereas business insurance for an electric bike is very cheap. For distances of less than a mile at a time, an electric bike can be used to deliver goods just as quickly and effectively as a small motorbike.

So long as the overall distance travelled in a single shift is less than 10–15 miles (16–24 km) and any one journey is less than one mile, an electric bike can be a very effective small delivery vehicle.

For longer distances, an electric motorbike can be an excellent alternative to a combustion engine. Electric motorbikes have several advantages over more conventional motorbikes.

- Lower running costs.

- No engine noise that might disturb people in a residential area.

- Excellent stop-start performance.

Some motorbikes, such as the excellent models in the Lexola range, include fast-charging capabilities as standard, making a genuine 24/7 use vehicle.

Is an electric vehicle suitable for my business?

In order to answer that question, there are various steps you need to take.

Evaluate the types of journey

For business use, electric vehicles are best used in a city or urban environment. Long-distance driving from one town to another is not so suitable for using an electric vehicle.

Electric commercial vehicles often have a limited top speed. This is fine for urban use but can be limiting for driving on freeways or across country.

The United Kingdom has a long history of electric commercial vehicle usage. The humble 'milk float' has been used from the 1920s to the present day for delivering milk and groceries to people's homes. This usage profile shows electric vehicles at their very best: lots of short journeys, frequent stops and a fixed route.

Electric commercial vehicles are exceptionally good in that particular environment, which is why they are particularly well suited to supermarket home deliveries, overnight parcel deliveries and postal applications.

Unless the overall range is limited, electric vehicles are not so useful for ad hoc work where the distance driven can vary wildly from day to day. If you know that the maximum distance the vehicle will have to travel is comfortably within the range of the vehicle, then this is not such an issue. However, look at the distance figures carefully to ensure this is the case.

Range

The biggest issue that businesses have with electric vehicles is range. All too often, a business will buy an electric vehicle in order to wave the green flag, only to find that their business requires them to travel further afield or on more individual trips than they anticipated.

The result is general frustration that the electric vehicle simply does not do what the business needs it to do. There is nothing more frustrating than having to turn down a customer because the electric vehicle is plugged in and needs to charge up.

Range fixation can also be an issue with employees using the electric vehicle. Drivers need to have the confidence that the vehicle will do what they need it to do.

It is vitally important that you spend time evaluating the distances that your current vehicles do every day. Once you know what range you need, look for an electric vehicle that is advertised as being able to do around double that range. This will cater for use in all conditions: with wipers, lights and heater or air conditioner running, driving with minimal consideration for fuel economy, or handling unexpected diversions.

So if you require a vehicle that can travel 25 miles (40 km) a day, look to buy one that can travel 50 miles (80 km).

If you have the opportunity to recharge your vehicle throughout the day, such as during lunch breaks, you can take this into account when working out a suitable range.

Realistically, if you require a range of more than 60 miles a day without recharging, an electric vehicle is probably not the right vehicle for you at the moment. You would be better waiting for another two years before considering an electric vehicle for your business.

Access to charging-points

If you are going to use an electric vehicle for business, you must ensure you are going to be able to charge it up when you need to.

In cities like London, where there is already a good public charging-point network in place, you may be able to charge your vehicles at these points. However, you cannot rely on them as other people will also be using these points. Easily accessible off-road parking with charging facilities is almost always essential if using an electric vehicle for business.

Charging facilities and commercial vehicles

If you are contemplating a larger electric vehicle, such as a large van or a heavy goods vehicle, you are going to require a high-voltage, high-power electricity supply in order to charge the batteries. In the UK, this means a 'three phase' 415 volt power supply, typically running at either 32 or 63 amps, while in the US a specific 'high-voltage/high-power' supply will need to be installed.

Electric vehicles and employees

Just like a new electric car owner getting fixated on range in their new electric car, an employee driving a works-provided electric vehicle also gets fixated on range.

If an employee believes the electric vehicle is not going to have enough range to complete their journey (even if the vehicle does complete the journey but the 'fuel gauge' is low), it can undermine confidence in the vehicle, which can spread amongst other employees rapidly.

If there is an electric car charging-point network in the town or city you are operating in, provide a satellite navigation system in the vehicle and program all the charging-point locations into the system. Even if the drivers never use the charging-points, just knowing they are there gives them the confidence to use the vehicle and know they are not going to be left stranded without power.

Promoting your business with electric vehicles

Electric vehicles make excellent promotional vehicles for a business. They attract attention and generate positive interest. If they are sign-written, they can generate enquiries for your business. There are few better ways of increase your reputation as an environmentally friendly company.

If you are looking to use an electric vehicle as part of your business and want to ensure that it also promotes your business, make sure it is well sign-written.

If your business is operating in an area where there are very few electric vehicles in use, make sure you also contact the trade press, local newspapers and radio stations. Local journalists are always looking for an interesting news angle and you will often get a lot of free publicity as a result.

Public charging-points

If your business is installing an outside power socket in order to charge up an electric vehicle, why not install two power sockets and make one available to other electric car owners?

If you run a shop or a restaurant, you will be encouraging electric car owners to use your business and if not, you can use the fact that you are installing an extra electric car charging-point as a great publicity tool to help promote your business.

Chapter summary

- Electric vehicles can make a lot of sense within a business environment.
- They provide economical transportation for inner-city use.
- They generate significant interest from both customers and the surrounding community, generating good publicity and goodwill.
- Ensuring an electric vehicle is a good 'fit' for your business requires careful evaluation of the way you use your existing vehicles.
- Electric vehicles can be a very effective way of promoting your business.

Appendix B: Electric Vehicles and Local Government

Governments, councils and public bodies are under pressure to reduce their carbon footprint. Implementing electric vehicles is a very positive way of demonstrating this.

As well as reducing the carbon footprint, electric vehicles are extremely visible to members of the public. They give the public a very clear sign of leadership that the government is making a stand.

Electric vehicles can very easily be used in many areas of local government and within public bodies. Park maintenance, city maintenance, shuttle bus services, internal mail delivery and community support are just a few of the ways that electric vehicles have been successfully implemented by public bodies.

How local government can help the adoption of electric vehicles in their locality

There are lots of ways that local government can help the adoption of electric vehicles in their locality.

Many of these can be carried out at little or no expense and in many countries there are central government subsidies available to assist in the more expensive tasks.

Free parking
Offering free or reduced-cost parking to electric car users is a simple way to encourage electric car ownership.

Allowing electric cars to use multi-occupancy lanes
Where cities have lanes dedicated to cars with at least one passenger, opening these lanes to owners of electric cars is another simple way to encourage electric car ownership.

Some councils have gone further and allowed electric cars access to bus lanes. While this may encourage early adoption of electric vehicles, it is likely to cause problems later. On balance, this is probably not to be recommended.

Electric car charging-points

Providing public charging-points is the single biggest thing a council can do to encourage electric car usage. Depending on how it is achieved, it can either be done on a strict budget or at great expense.

If a council is planning to implement their own electric vehicles, they are going to need to install charging-points for these vehicles.

These charging-points are typically an outside power socket, fitted to a wall. If councils are installing these, they should consider installing them in publically accessible areas and allow other electric car owners access to these charging-points.

Installing power supplies in multi-storey car parks, where the power already exists, is a relatively straightforward and cheap way to provide public charging-points. The spaces must be reserved for electric car owners who are charging their vehicles and this must be enforced.

While councils can choose to install proper electric car charging-points in car parks, this is not actually necessary. All that is required is an industrial quality 13 amp socket.

If cost is a concern, councils can charge electric car users for electricity use through an additional tariff on the car park ticket. Many car park ticket machines already have the ability to add ad hoc tariffs in this way.

Many councils have expressed concerns that these charging-points might be damaged or deliberately abused. Although this has been a valid concern, charging posts have been installed across the UK for several years and vandalism has not been an issue: The City of London and the City of Westminster, who installed their first charging-points in 2000 and 2001 respectively, say they have not experienced these problems.

On-street charging-points are another option but are significantly more expensive to install. Companies like Chargemaster and Pod Point in the United Kingdom and

GreenlightAC in the United States all supply and install on-street electric car charging-posts.

Incidentally, for electric car charging-points to become viable, they must be available to all electric car owners and not just those who live in the area the charging-points are being installed in.

In the past, some councils have installed charging-points but have then barred any electric car owners except those living in the region from using them. This pointless exercise causes a lot of frustration and loss of goodwill amongst electric vehicle owners. Local owners are able to charge from home and do not need charging-points within a mile or two of their house. Meanwhile, other electric car owners are unable to use the facilities and therefore are less likely to travel to your area.

Encouraging local businesses

'At Work' and 'Gone Shopping' charging-points are also highly beneficial for some electric car owners. They certainly help encourage electric car ownership. Encouraging local businesses to offer these can benefit their own businesses as well as their employees and the community at large.

Request that electric car charging-points are incorporated into new designs when private enterprises are submitting plans for retail parks or industrial areas.

Charging at home

Many people who would like to be able to own an electric car do not have off-road parking available to them. Offering a paid-for service to install a power bollard and reserve a space outside the owner's home, at their own expense, will encourage a number of people who currently have nowhere to charge up an electric car to buy one.

Do people use electric cars instead of bus services?

One concern that many councils have is that encouraging electric cars will take people off the buses and back into their own cars again.

The G-Wiz Electric Car Owner's Club ran a poll amongst its London-based members in 2007, asking the membership if they bought an electric car instead of using public transport. Every single member who responded to the survey said this was not the case.

Instead, the electric car replaced a conventional car. Somewhat surprisingly, a number of SUV owners have been amongst the earliest adopters of electric vehicles and so electric cars are quite often replacing a 4WD vehicle when travelling into town.

Other than this arbitrary questionnaire, very little research has been carried out into this. The City of London used this concern as a reason to remove free parking for electric cars from their car parks, yet admitted at the time that they had carried out no definitive research into this.

At present, the best that can be said is that it appears unlikely that people buy electric cars as an alternative to using public transport.

Chapter summary

- Electric vehicles can work well in a number of applications within local government and public bodies.

- If you are installing charging-points for your own electric vehicles, why not install some of them in publicly available locations and offer them to other electric car users?

- Local government has a part to play in encouraging electric car ownership within the community. Start with the simple, low-cost options and build from there.

- Encouraging local businesses to play their part is also important.

- Offer charging options for people who have no off-road parking. You will not be able to help everyone but you will be able to help some.

- There is no evidence to suggest that people use electric cars instead of public transport.

Appendix C: Other Electric Vehicles

Electric vehicles are not limited to electric cars. Electric bikes, motorbikes and commercial vehicles have all been available for a number of years.

Many industry observers, myself included, believe that, while electric cars will become more mainstream over the next few years, the real revolution in electric vehicles will be in an entirely new form of personal transportation: a product that is not a car, nor a bike, but an entirely new vehicle for urban mobility.

The future of transport?

Today, we live in a world where we love our cars. We love the freedom that they give us, the flexibility to go wherever we want to go, whenever we want to do it.

Yet much of that belief is a myth. Traffic and congestion makes driving in our cities stressful. Finding car parking spaces in cities is often difficult and expensive. Stop-start driving in towns and cities can often mean travelling at the same average speed as walking. In many parts of the world, travelling by train is much faster, safer, cheaper and often more convenient than long-distance driving by car.

The ever-increasing cost of running a car, with higher insurance prices, fuel costs and servicing costs, is also an issue. More and more people, especially young people, can no longer afford to own and run a car.

Furthermore, young people no longer view cars as freedom. Whereas people over the age of thirty remember owning their first car as their first step of freedom, younger people see freedom as accessing Facebook from their mobile phones. Instead of meeting people face to face, young people meet online and the role of a car as a way of escaping from parents and meeting friends is much less important.

So it is not a surprise that the vehicle of the future may not actually be a car at all. If an electric car is just an ordinary car with an electric motor and batteries, congestion in our cities is not going to get any better. And while the fuel cost will be significantly less, the cost of owning a car is still going to be high. So the question is, what will this vehicle of the future actually be?

Personal mobility – using an electric vehicle to get from A to B

Next time you are stuck in a traffic jam, count how many cars you see where there is only one person in the car. If you are stuck in traffic commuting to work, the chances are that almost all the cars around you will have a single occupant.

So what if a new form of transport existed that allowed one person to make that journey? A vehicle that could not be called a car, but could provide a simple and effective way for one person to get from A to B on a daily basis?

Of course, such a vehicle wouldn't have the flexibility of a car. But because it would be smaller, simpler and more efficient, it could also be significantly cheaper to buy and run. It could possibly be so cheap to buy and run that it would be cost-effective for existing car owners to buy one *alongside* their existing car.

Of course, such a vehicle would not suit everyone. It wouldn't be suitable for longer or higher-speed journeys. It would be really designed for driving through a city or urban environment, where traffic congestion is a real problem and getting around by car is difficult.

I'm going to let you in to the biggest secret of the electric vehicle industry today. It is so secret that most of the people who work in the industry, most of the designers, most of the automotive companies and almost all of the supposed industry experts and consultants do not know it. This electric vehicle already exists in its embryonic form. Not only that, but worldwide sales have been increasing significantly over the past decade – from an estimated ten million a year in 2005 to an estimated fifty *four million* in 2014. An estimated 200,000,000 are now in regular use around the world.

The vehicle in question is the electric bicycle. Small, simple and easy to use, the electric bicycle replaces short car journeys with a bike ride. The rider has the choice of using the electric motor assistance or pedalling like an ordinary bike. In cities such as London, an electric bike is often the fastest way of getting through the city. Throughout the world, a staggering number of people are buying and using electric bikes as a way of getting from A to B simply and effectively.

Sales of bicycles in general have been increasing over the past few years. In Europe, more bicycles are sold each year than cars: in the UK, for example, bicycles outsell cars 3:2. In the United States, bicycles also outsell cars with nearly four million more bicycles being sold each year than cars.

Electric bicycles have created an interesting paradox, however. A number of years ago, I ran a company selling electric bicycles and I found out an interesting fact. Cyclists do not use electric bicycles: they find them too heavy and cumbersome. Electric bike owners don't want to cycle and would really prefer a vehicle with a bit more practicality. In other words, a huge industry has been created around a product that people do not actually want!

A very small number of people currently in the electric vehicle industry are just starting to see what this could mean. If 200,000,000 people around the world have bought an electric bike because it is *almost but not quite* what they want, there is a huge market for a product that is absolutely what people want to buy. The question is *what is this product?* Because nobody has ever created such a product, nobody is entirely sure what that product needs to be.

The first tentative steps

That has not stopped people and companies from trying to create the ideal personal mobility vehicle. The most famous of these is the Renault Twizy, the first attempt by a major car maker to try to understand this market.

The Renault Twizy is now on sale across Europe. It is technically a quadricycle, and is the first attempt by a major car manufacturer to try and create a genuinely new type of road-going vehicle, as opposed to just putting an electric motor into a car.

The result is a four-wheel motorbike, with seating for two people, one behind the other, a roof, optional doors, and safety equipment such as airbags and crumple zones.

Because the Twizy does not pretend to be a car, or have all the same features as a car, it is extremely compact, lightweight and cheap to buy and run. Whereas most electric cars are expensive to buy, Twizy is one of the cheapest four-wheel vehicles on the market. In many countries, people can lease a Twizy for less than the cost of two tanks of fuel per month.

Across Europe, light quadricycles such as the Twizy can be driven by anybody over the age of sixteen. In Fance, Italy and Belgium, you can even drive one without having a driving licence.

The Renault Twizy

It's not perfect. Nor will it suit everybody. But that is not the point. The point is that there is something different happening to personal transport. Other manufacturers are working on vehicles like these. Given time, it is something that could radically change the way we think about cars and how we use them in the future.

Other alternatives available today

If a Renault Twizy isn't the right vehicle for you, there are plenty of other options that are available for you to buy right now.

Electric bikes

An estimated 54 million electric bicycles were sold worldwide in 2014, up from 27 million electric bicycles in 2010. In almost every country in the world, electric bicycle sales is one of the fastest-growing consumer markets. In countries like Germany, the United Kingdom, Sweden, the Netherlands and France, electric bicycles are a very common sight. If you live in one of those countries and you haven't noticed one, it is simply because they look very much like ordinary bikes to the casual observer.

Without doubt the most popular electric vehicle around, an electric bicycle is simply a pedal cycle with electrical assistance. In Europe and the United States, they are regarded in law as pedal cycles and can be used by anyone aged fourteen or over, without road tax or insurance.

Electric bicycles have gained popularity with commuters and with older people who still want the freedom of a bicycle without the hard work of cycling up the hills.

Top speed is typically 15 mph (25 km/h) and the range is usually around 15–25 miles (25–40km). The batteries can usually be detached from the bike and taken inside for charging. Some customers choose to buy two battery packs so they always have one charged up and ready to go.

A budget electric bicycle can be bought from around £400 in the UK, or $500 in the US, with quality makes starting from around £900 in the UK and $1,200 in the US. Urban Mover is one of the leading manufacturers of electric bikes.

Electric motorbikes and scooters

Electric motorbikes and scooters are also gaining in popularity, predominantly as commuter vehicles.

If you have only driven a car in the past, you may need to take another test before being allowed to drive a motorcycle. Legislation varies depending on the power of the motorcycle, when you took your original driving test and which country you live in.

Most electric motorbikes are scooter-styled bikes, with a range of 40–50 miles (65–80 km). Top speed varies depending on the power required but is typically between 30 mph (45 km/h) and 70 mph (112 km/h).

Prices for electric motorbikes are comparable to the equivalent combustion engine motorbikes. Zero and Brammo are the two leading manufacturers of electric motorbikes and Peugeot and Honda are the two leading manufacturers of electric scooters.

Electric commercial vehicles

There is a separate appendix on electric vehicles in business that covers this area in more detail, but a number of electric commercial vehicles are already available on the market.

Most electric vans have a range of between 50 and 100 miles (80–160 km). At present, many of them are expensive to buy, although the price differential is dropping quickly. Much of this additional cost is offset by the significantly lower running cost of the vehicles, especially if spread over a five-year ownership.

Electric boats

Small electric boats for inland waterways have become popular as recreational craft over the past ten years.

They are popular because they are silent and because they are cheap to buy and run. You can buy an electric-powered outboard motor and battery for around half the price of an equivalent combustion engine outboard motor with fuel tank.

As electric boats are silent they do not disturb wildlife, which makes exploring rivers and lakes particularly enjoyable.

The motors do not need to be excessively big or powerful, so even a modest-sized leisure battery, as used in caravans and RVs, can provide eight hours or more of power. Many people have combined their electric boat with a solar panel so the boat can recharge when it is not in use.

Chapter summary

- Electric vehicles have quietly been making inroads into the market for the past ten years.
- Even if an electric car is not a practical solution for you now, there are other electric-powered vehicles available that are suitable for recreational or utility use.

Appendix D: Charging an Electric Car with Renewable Power

Many people who buy an electric car want to reduce their impact on the environment. So it is probably not a surprise that many of them decide they want to charge their cars up using a renewable energy source, such as solar or wind power.

Charging up an electric car with solar

If you want to charge up an electric car with solar power, it helps if you live in a sunny climate with a good amount of sun throughout the year. If you live in California or Italy, for example, it is possible to use sun power to charge and run your electric car all year round.

Elsewhere, in northern Europe or Canada, for example, you are unlikely to be able to generate enough power in the winter months to make your car fully solar-powered.

The easiest solution for charging up an electric car with solar power is to build a grid-tied solar array. This generates electricity that can then be used to charge up the car, run the family home or be sold back to the utility companies if the array is generating electricity when you do not need it.

Costs for installing solar power are falling, but do not expect a solar charging system to be cheap. Even in sunny climates you are likely to require a 2 kWh solar array, at least, in order to charge up an electric car on a daily basis, at a likely cost of £3,000–£5,000 in the UK, or $5,000–$8000 in the United States.

Of course, if you need a shorter daily range, you can use a smaller solar array. A company called Alpha Energy in Phoenix, Arizona, has recently launched a new scalable solar electric charger called the EV-500. This unit is specifically designed for charging up 48 volt electric vehicles, such as many smaller electric city cars, NEVs and electric scooters.

Solar-powered cars

A few specialist manufacturers have announced solar-powered cars. Venturi and Mahindra both showed prototype vehicles and announced they would be put on sale a few years ago, but since then nothing has appeared.

Solar-powered cars are viable for compact, lightweight cars, particularly those travelling at low speeds such as within towns and cities, with frequent stops. Solar can be used to provide a top-up charge and would be unlikely to provide the sole power source. Purely on solar power and in a sunny climate, sunlight could provide a solar-powered range of around 5–15 miles (8–24 km) per day.

Although the range may not seem that great, there are many drivers who live in a sunny climate and only use their cars for short journeys a few times each week. For these people, they could now drive an entirely solar-powered car.

Below: The Venturi Eclectic was a concept city car for Europe that was said to have a range of up to 8km (around five miles) per day on solar power. Originally planned for production in 2011, production was eventually shelved.

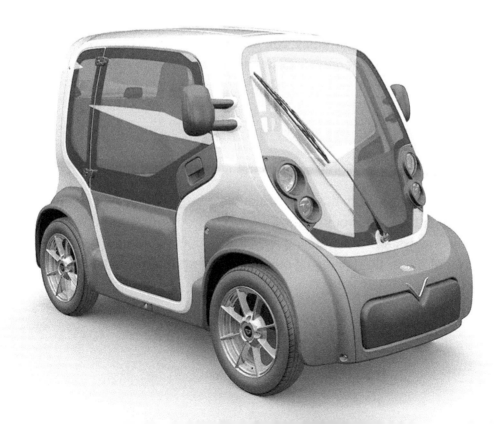

Over the past three years, I have had serious discussions with electric vehicle designers and decision makers working for mainstream car companies about the potential for solar-powered cars. The conclusion of these meetings has been that solar-powered electric cars are inevitable. It is just a matter of time before a mainstream manufacturer builds one.

These cars are not going to be suitable for everyone. Yet the first exciting steps towards practical solar road cars are being made behind the scenes. With the advancement of solar panels with better capacities and lower costs and the ongoing development of electric cars, it may not be that long before solar electric cars become a common sight on our roads.

Solar power at home

Many electric car owners have now installed solar power at their homes, using the solar power to charge their cars from sunlight.

From a pure cost point of view, charging your car from your own solar power station at home is inevitably more expensive than paying for your electricity, particularly as an electric car can consume quite a significant amount of energy, particularly if you are regularly travelling long distances with your car. However, it is satisfying to know that your environmentally friendly car is being powered from an environmentally friendly source and being 'energy independent' does bring a level of security for many owners.

Commercially available solar charging-stations

The first commercially available solar charging-stations are now appearing.

Beautiful Earth, a New York based sustainable energy company, has unveiled its first solar charging-station in Brooklyn. Powered purely on sunlight and with no mains grid connection, the system is built using recycled shipping containers and can charge up an electric car faster than plugging one into a standard power socket.

Meanwhile, in the United Kingdom, the Cross Group has recently unveiled a commercial multi-bay solar-powered electric vehicle charger. Designed for businesses and local government, the system allows multiple charging-points to be installed anywhere, without requiring a mains power connection.

At a cost of around £30,000 (approximately $48,000), it is comparable in price to installing multiple street level charging-points on a busy street and similar to the cost of a rapid charger installation. The benefits are that the charging stations can be installed in places such as laybys where it would be impractical to provide power in any other way.

The technology, the products and the potential to have a solar-powered electric car charging network covering an entire country now exist. Surely, it can only be a matter of time before a country implements one?

Finding out more about solar

If you are interested in finding out more about solar power, I would recommend visiting *www.SolarElectricityHandbook.com* to find out more about the technology. The website includes a solar project analysis tool that allows you to estimate the size and cost of a suitable solar implementation.

Wind power

Wind turbines are gaining popularity, both as a way of generating large amounts of power for the utility grid and also as a way of generating small-scale energy for a household.

Wind power is nothing new. Back in the 1920s and 1930s, wind turbines were used by many farms in order to provide electricity. Their use died out as the utility grids expanded, providing cheap, dependable electricity at the flick of a switch.

Today, large-scale wind turbines are efficient and effective. They can be installed in a variety of locations comparatively quickly. Unlike the early large wind turbines, modern turbines are virtually silent and the largest systems can generate between two and six megawatts of power.

Small wind turbines do have disadvantages, however, and are very site-specific. Compared to the large wind turbines used by the power companies, small wind turbines are not particularly efficient and need to be situated in an area of above average wind in order to generate reasonable amounts of power.

Wind turbines also require a very smooth airflow. Smaller turbines are very susceptible to turbulence. If you live near a busy road, near trees, or in a built-up area, a wind turbine is unlikely to work well for you. Turbulent air (where the wind is constantly changing direction) leaves the wind turbines constantly chasing the wind direction rather than extracting power from it.

If you have the right location, wind power may work for you. Wind turbines work best in open, exposed areas where average wind speeds are high.

Ideally, wind turbines should be mounted high up. Even small wind turbines are often mounted 10 metres high in order to get sufficient wind power. Turbines should also be installed away from buildings. Roof-mounting a wind turbine is not ideal, as the building itself generates turbulence and the vibrations from a wind turbine being affected by turbulent air can be magnified through brick walls, creating a loud and annoying vibrating sound.

Small wind turbines can often be found on small leisure boats. The turbine maintains the batteries when the boats are not in use. Boat owners have mixed reports on their benefits.

- Owners who have the turbines mounted relatively low down on the boat find they do not perform well at all.

- Owners who have the turbines mounted high up on a mast report much better performance.

Fitting a wind turbine onto an electric car

I have been asked on a number of occasions whether it would be beneficial to fit a wind turbine onto an electric car.

Fitting a turbine to a car and using it while driving the car would not work. The amount of additional drag created by the wind resistance would outweigh the benefit of electricity generation from the turbine.

Fitting a turbine to a parked car in order to recharge the batteries is a possibility, but is unlikely to yield significant results. Electric cars are typically used in built-up areas with poor wind flow, so unless the car is parked in an open, windy location, the amount of power generated by the turbines is likely to be very small.

Chapter summary

- Many people buy an electric car in order to reduce their impact on the environment. These people can reduce their impact further by using renewable power.

- If you have solar panels on your home, you can use solar energy to charge up your electric car during the daytime. This is more feasible in sunny climates, such as the southern states of America, than it is in Canada or northern Europe.

- You can buy a solar charger kit specifically designed to charge up an electric car. Again, this is more feasible in sunnier climates.

- Cars with solar panels mounted on the roof will start to become available. In the right climate, short-distance drivers may find that a solar car provides all the power they ever need.

- The technology and products required to build a nationwide network of electric car solar charging-stations already exists.

- A wind turbine could be used to generate electricity for charging an electric car, but a turbine that is mounted to the car when it is parked is likely to be very inefficient and not produce enough power to make it worthwhile.

Appendix E: Free – Working Towards a Radical Price

I am often asked whether electricity will become the dominant fuel for cars in the next few years. I believe it will, but the timescale for this switch is unclear. Much will depend on the cars that become available and how quickly consumers see the benefits and advantages of electric cars.

Many industry commentators believe that by 2020 only 2 or 3% of the cars on our roads will be electric. The CEO of Renault-Nissan, Carlos Ghosn, believes the take-up will be much faster, with 10% of all new cars being electric-powered by 2020. Ford goes further still: Derrick Kuzak, Ford Group's Vice President of Global Product Development has said that electric drive models could account for a quarter of Ford's global automotive sales by 2020.

Undoubtedly there will be a few incidents along the way and, just as undoubtedly, some cars will be better than others. Yet I personally believe that, at some point, electricity will become the dominant fuel for cars and that the take-up could be much quicker than currently anticipated.

My reasoning is one of simple economics. Oil prices are only ever going to go up. Meanwhile, the price of an electric car is going to fall. Eventually, the cost of leasing an electric car will be cheaper than the cost of putting fuel into a conventional car.

At that point, the cost of the electric car itself is effectively zero.

Anyone want a free car?

Throughout the world, top economists have been talking about how business and buying patterns will evolve over the next thirty years.

There is consensus among these economists that the real cost of many products and services will drop significantly, to the point where many of the products and services we currently pay for become free.

In his book, *Free: the Future of a Radical Price*[76], Chris Anderson argues that the economics of abundance forces the devaluation of products and services to the point where they are virtually free. Zero pricing is changing the face of business. Chris argues that businesses and industries have the choice of either adopting this strategy for themselves or becoming victims of it.

For a product to be free, there has to be a related product or service that can be charged for. The free product must also be low-maintenance and have negligible ongoing costs associated with it.

The mobile phone model

A good example of a free product that we all have and use are cell phones. If you walk into a cell phone shop, almost all the phones are offered free of charge. You simply choose the model, select your usage tariff and off you go. If you are an existing customer with an older phone, you can upgrade your phone and still not pay anything for your new purchase.

The actual manufacturing cost of the phones is most certainly not zero. Many phones actually cost hundreds for the phone companies to buy. The profits made on monthly usage charges are used to write off the cost of the phone itself.

Making the mobile phone sales model work for cars

By removing the dependency on oil, this sales model will be able to work for electric cars in the future. The cars themselves will be free and customers will choose a usage tariff that suits them.

---

[76] *Free: The Future of a Radical Price*. Published by Cornerstone Digital. 15th December, 2010.

Usage tariffs will directly replace the fuel bills that everyone has to pay to use their existing cars. They will be calculated to work out at either the same cost or slightly cheaper than putting fuel into an equivalent combustion engine car.

So instead of paying £250 or $250 per month for fuel at service stations, you would pay a similar amount each month for the usage plan on your electric car. The electric car itself would be free. At the start of your contract, you'd be able to go into the car showroom, choose the car that you want and simply drive it away.

Given the choice between buying a $15,000 car with a combustion engine and then paying $250 per month on fuel, or just paying the $250 a month on a usage plan and getting the equivalent electric car for free, which would you choose?

Economically, we're really close to this working already in the United Kingdom. As I explained on page 58 in the section *Reaching the Tipping Point*, I used to pay £250 a month for the fuel in my old car. I now pay £220 a month to lease an electric car and the additional electricity bill is only £17 a month. So other than my initial payment of £1,800 when I first collected the car, I am already effectively getting a free car.

The figures don't yet work for everyone: you have to be driving at least 15,000 miles a year and based in Europe where fuel prices are higher than the United States. Yet that already covers a large proportion of the car buying public and as electric cars become cheaper and fuel costs become more expensive, the calculations are going to start working for even more people.

So why aren't electric cars already being sold like this? It's simply a case of presentation and marketing. It took the mobile phone industry around fifteen years to get to the point where phones were given away free of charge, but the car industry has got to the same point, just four years after the first mainstream manufacturers launched their first electric models.

Expect the change to happen and expect it to happen soon, particularly in Europe, where the cost of fuel is so high. Walking into your local car dealership will be like walking into your local mobile phone store. Choose your electric car and drive it away.

Of course, this model will only work with electric cars. With hydrogen cars or petrol/gasoline or diesel cars, you still have to pay for fuel so there is no cost saving available to customers. For this reason and this reason alone, electric cars will become the dominant type of vehicle on our roads.

Why would anyone choose anything different?

Index

Aerodynamics
    Wind Resistance 39, 215
Air Pollution 128, 129, 136, 137, 138, 151, 156
    Asthma 136, 137
Aixam
    Mega City 173, 174
Asthma .... *See* Air Pollution
Autonomous Vehicles 31
Batteries
    Environmental Impact 127, 176
    Guarantees 45
    Lead Acid 176, 177, 178
    Lithium ion 179
    Nickel Metal Hydride 176
Battery
    Replacement 182
BMW 21
    i3 8, 16, 34, 35, 54, 56, 74, 87, 93, 173, 174
Braking 35
Car Club 58
Car hire 58
Car Sharing 59
Carbon Footprint ....
    126, 148, 155, 156, 170, 171, 174, 178, 179, 180, 182, 186, 189, 200
    Dust to Dust 181
Cargo Trike 30
Charging at work 41, 57
Charging point networks 76
Charging Points 11, 42, 57, 197, 198, 213

At Home 52, 202
Caravan Sites 41
CCS 70, 71, 72, 74, 85, 86, 87, 89, 90, 93, 109
CHAdeMO 70, 71, 72, 74, 77
Elektromotive 201
GreenlightAC 202
Home charging 54
Hotel 41
How to use a public charging point 75
Independent Businesses ....
    .... 42, 54, 199, 202
Nationwide Charging Infrastructure
    .... 1, 44, 169
Pod Point 201
Power Bollard 53, 202
Pubs 41
Restaurants 41
Shops 42, 202
Chevrolet 175
    Spark EV 55, 69, 90
Citroen
    C-Zero 51, 74, 91
Coal 139, 148, 150, 164
Combustion engine ....
    7, 15, 16, 18, 19, 20, 22, 39, 61, 65, 126, 127, 168, 170, 174, 208, 209, 218
Crude oil 143, 144
Diesel Particulate Matter 131, 132
Drax Power Station 152
Ecotricity 76

Electric Bicycle..................30, 195, 208
Electric Boat 209
Electric Bus.........................195
Electric minibus..................195
Electric Motorbike 195, 196, 208
Electric Truck.......................195
Electric Van............................195, 209
Elektromotive ....See Charging Points
Emissions19, 21, 126, 127, 130, 131, 132, 134, 135, 138, 140, 147, 150, 152, 153, 155, 157, 164, 165, 170, 172, 173, 176, 179
   Biodiesel 133, 134
   Carbon Monoxide130, 132, 134, 135, 151
   Diesel 19, 131
   Electricity Production 140
   Nitrogen Oxide130, 132, 135, 151, 156
   Sulphur Dioxide 135, 151
   Tank to Wheel..............................
   .... 126, 135, 174, 175, 188, 190
   Well to Wheel..............................
   126, 142, 147, 171, 175, 176, 190
Environment ....
   19, 40, 48, 125, 194, 196, 199, 210, 215
environmental .... See Environment
Fast Charging 7, 11
FIAT
   500e 92
Ford..175
   Focus Electric 54, 74, 93
Free Car 216
Fuel Cell 2, 20, 22

Fuel Economy.....................................
   48, 127, 131, 171, 184, 197, See Range
Gas fired power station 155, 156
Geo-thermal energy 139, 157, 164
GreenlightAC .... See Charging Points
Honda 19, 180
Hybrid 2, 15, 16, 18, 19, 169
Hydro-electricity 157
Hydrogen ....See Fuel Cell
Hydrogen Fuel Stations 20
KIA
   Soul EV 74, 94
Leasing 61
Low Speed Vehicles.......................... 81
LSV .... See Low Speed Vehicles
Mahindra
   e2o 95
Medium Speed Vehicle..................... 80
Mercedes...181
   B-Class Electric Drive................... 96
Milk Float...196
MINI 175
Mitsubishi............. 173, 174, 176, 186
   i-MiEV34, 35, 51, 52, 55, 56, 69, 74, 91, 98, 110, 118, 119, 121, 173, 174, 186, 188, 189, 191
   Outlander PHEV 15, 17, 18
Mobile Phone 218
National Grid 139, 164, 170
Neighborhood Electric Vehicle 38
   NEV 38, 80, 81
NEV . See Neighborhood Electric Vehicle
Nissan
   e-NV200 .... 74, 97, 100, 195
   LEAF ....
   5, 8, 14, 15, 34, 35, 40, 51, 52, 54, 56, 58, 61, 65, 69, 72, 74, 88,

93, 96, 97, 98, 99, 102, 109, 111, 120, 173, 174
Nuclear Power 148, 156
Oil power station 156
PACTS.. 14
Performance2, 33, 34, 38, 78, 128, 146, 153, 170, 214
Peugeot
   iOn 34, 51, 61, 74, 91
Phoenix 210
Pod Point..................*See* Charging Points
Pollution ....
   19, 23, 29, 45, 113, 125, 126, 127, 128, 129, 130, 131, 134, 136, 137, 138, 139, 140, 152, 153, 156, 161, 182
   Diesel 19, 132
Pumped Storage.....................158, 159
Purchasing......................... 61, 66, 217
Quadricycle 82, 83
Range...
   9, 10, 15, 18, 22, 35, 36, 38, 39, 40, 44, 45, 49, 50, 57, 60, 130, 196, 197, 198, 208, 209, 210, 211
   EPA..
      50, 51, 52, 85, 86, 87, 89, 90, 91, 92, 93, 94, 95, 96, 97, 98, 100, 101, 102, 103, 105, 106, 108, 109
   NEDC...
      50, 51, 52, 85, 86, 87, 89, 90, 91, 92, 93, 94, 95, 96, 97, 98, 100, 101, 102, 103, 105, 106, 108, 109

Range Fixation 35
Recycling128, 153, 178, 181, 182, 212
Regenerative Braking 18, 35, 40
Renault 216
   Fluence 93, 114
   Kangoo 100, 195
   Twizy 24, 78, 101, 206, 207
   Zoe 34, 69, 102, 173, 174
Reva 180, 186
   G-Wiz 186, 203
Road Noise 13, 33
Running Costs 48
self-driving vehicles*See* Autonomous Vehicles
Servicing 65, 66
Silence 13, 14, 33, 209, 214
Smart 61, 175
Smart Meter 170
Solar ....
   160, 209, 210, 211, 212, 213, 215
Tesla
   Model S ....
      74, 103, 104, 105, 116, 117, 118
   Model X.......................................105
   Roadster 35, 108, 118
Toxic Organic Micro Pollutants 130
Toyota 15, 175, 176
VW
   e-Golf 56, 74, 109
   e-Up!..109
Website 170, 174, 193
Wind Noise 13, 33
Wind Turbine 159, 160, 213, 214

Lightning Source UK Ltd.
Milton Keynes UK
UKHW030045070219
336888UK00005B/523/P